The Last Volcano

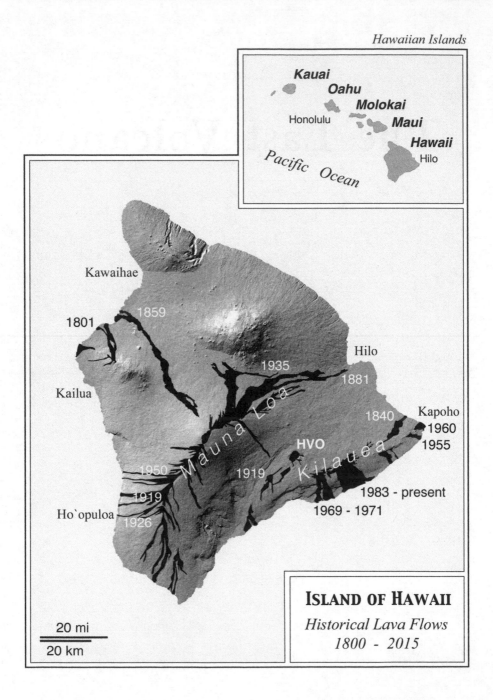

Hawaiian Islands

Kauai

Oahu

Molokai

Honolulu

Maui

Hawaii

Pacific Ocean

Hilo

Kawaihae

1801

1859

Hilo

1935

1881

Kailua

Kapoho

1840

1960

HVO

1955

1950

1919

Kilauea

Mauna Loa

1919

1983 - present

Ho`opuloa

1926

1969 - 1971

20 mi

20 km

ISLAND OF HAWAII

Historical Lava Flows
1800 - 2015

The Last Volcano

A MAN, A ROMANCE, AND THE QUEST TO
UNDERSTAND NATURE'S MOST MAGNIFICENT FURY

JOHN DVORAK

PEGASUS BOOKS
NEW YORK LONDON

THE LAST VOLCANO

Pegasus Books LLC
80 Broad Street, 5th Floor
New York, NY 10004

Copyright © John Dvorak 2015

First Pegasus Books edition December 2015

Interior design by Maria Fernandez

FRONTISPIECE: Shaded relief map of the island of Hawaii showing historical lava flows, 1800–2015. The years of some major flows are indicated. HVO indicates the location of the Hawaiian Volcano Observatory at the summit of Kilauea volcano. *Modified from a map provided by Katharine Cashman.*

Library of Congress Cataloging-in-Publication Data is available.

ISBN: 978-1-60598-921-1

10 9 8 7 6 5 4 3 2 1

Printed in the United States of America
Distributed by W. W. Norton & Company

To Joyce and Sarah,
who taught me the important lessons of life

CONTENTS

NOTE TO THE READER

I have dispensed with diacritical marks in most Hawaiian words. For example, I use "Hawaii" rather than "Hawai'i." But when it was necessary to distinguish clearly which Hawaiian word I meant—or to aid the reader in pronunciation—I have used them. An important example is "Halema'uma'u," which is the name of the large crater at the summit of Kilauea volcano.

NOTE TO THE READER

They yearn for what they fear.

Dante, *The Inferno*

The Last Volcano

A CITY HAS PERISHED

I n the spring of 1902, in the tranquil light of a tropical sunset, a ship, seemingly without purpose or crew, drifted into Castries harbor on the small Caribbean island of St. Lucia.

The masts and rigging were smashed. The normally crisp white sheets and awnings were hanging loose, torn and burned. A rescue party was organized and several of St. Lucia's leading citizens were rowed out in a large boat to investigate. One of the investigators was Charles Dennehy, the British colonial surgeon on St. Lucia.

As Dennehy later told the story, as soon as he stood on deck, all sense of reality gave way. The entire ship was blanketed with a thick layer of fine gray ash. In places, the ash had formed into drifts. And beneath those drifts, as Dennehy and the others soon discovered, were the bodies of dead men.

Four dead bodies were found on the afterdeck, five more forward. Each one was in a highly contorted state, arms and legs at odd angles, an indication that each man had died in great agony. The faces of the poor victims had all but disappeared, scorched away by intense heat. Only a few patches of flesh and hair still clung to the skulls, making the men unidentifiable.

The body of one man was found intact, his position so natural that, at first, Dennehy thought the man must be asleep. He would later be identified as the chief engineer. He was found sitting in a chair next to the engine room, his head tilted to one side, his hands resting naturally on his lap.

As Dennehy continued to search, he heard faint noises coming from below. He descended a ladder and entered a dark room. Inside were a dozen men, still alive. Some were lying on bunks. Others were leaning against walls. The head and clothes of each one were covered with the same fine ash that blanketed the ship. Most of the men were silent. A few gave out pitiful groans. As Dennehy began to examine the nearest man, another one rose and approached him, moving slowly, as Dennehy recalled, "as if suddenly struck by the feebleness of old age."

"You don't know me?" the man asked.

Dennehy studied the figure. The face was blackened and blistered. The arms and hands were bloodied and swollen to three times their normal size. The man repeated the question. This time Dennehy recognized the voice.

"Captain Freeman, what has happened? Where have you come from?"

"From the gates of hell," he replied.

Captain Edward Freeman, four years in command of the *Roddam*, a steamship that regularly carried cargo and passengers between London and the Caribbean, had left St. Lucia the previous night. He had sailed his ship through a raging storm, wishing several times he had never left. But Freeman was anxious to know what had happened on Martinique. Four days earlier, in Barbados, he heard that the volcano on Martinique, Mount Pelée, was erupting.

Two days later, in St. Lucia, he was told that an earthquake beneath Martinique had severed an undersea cable, ending all communication with the island.

By morning the weather had cleared and Martinique was in sight. Freeman could see the island's main city, St. Pierre, stretched out for two miles along the coastline. Twenty-six thousand people lived in St. Pierre. Another 160,000 lived elsewhere on the small island. And behind the city loomed Mount Pelée, a plume of light gray ash rising from the summit.

The eruption had started the previous month, the volcano spewing ash into the air. By May 8, the day the *Roddam* arrived, several inches of the powdery material had fallen over St. Pierre, giving the normally festive city an oppressive look.

A dozen ships were in port. Each one was tied to a buoy and floating near shore. Freeman recognized the *Grappler* of the West Indian and Panama Telegraph Company where his brother-in-law served as third mate. The *Grappler*'s crew was already awake and setting out equipment, probably preparing to find and repair the severed undersea cable. Later, when asked why he had entered the port if a plume of ash was rising from the volcano, Freeman would answer that he had seen other ships. "An indication," he explained, "that it was safe as they would know the situation better than I would."

He saw the harbormaster on shore signaling to him with flags to anchor at the south end of the city. He stopped the *Roddam* a few hundred yards from shore, then ordered the crew to secure the ship to a nearby buoy with an iron chain. From this close vantage point, he again studied the city. "How pretty and gay St. Pierre managed to look in spite of ash," he would remember. "The Cathedral glistened in the morning sun. Then from the towers came the sounds of bells calling the faithful to morning Mass."

Freeman could see people hurrying between shops. Others were hanging colorful flags and banners between large buildings and across main streets. It was Ascension Day and the people of St. Pierre were preparing their city for a celebration. Normally, he would have ordered the cargo unloaded immediately, but, because it was a holiday, he decided he would let his crew rest and remain at St. Pierre one day.

An hour after the *Roddam* arrived, the ship's agent, a Mr. Plissoneau, came alongside in a small boat. He was climbing an outside ladder and preparing to board the *Roddam* when the volcano let loose.

Freeman was on the bridge in the map room. He heard a muffled roar. He leaned out an open doorway and looked at the volcano. He could see a small black cloud begin to billow up from the summit, then spill down the side of the volcano and toward St. Pierre. As the cloud began slide, Freeman called down to the agent. "Come up here. You can see better."

In silence, the two men watched as the black cloud approached, holding close to the ground, Freeman describing it as "an awful thing—a fascinating thing." The front of the cloud was crossed with short bursts of lightning. Occasionally, red glowing rocks were thrown out in front of it. Freeman felt no disturbance in the air. Only a faint rumble could be heard coming from the volcano.

It took two minutes for the cloud to travel the four miles and reach St. Pierre. By then, the front was a mile wide and rose more than a thousand feet into the air. Freeman now realized it was racing at him at hurricane speed.

"Take shelter!" he yelled as he jumped back into the map room, diving beneath a stack of empty sacks. Within seconds, the cloud hit the water. As it did, it caused the sea to well up and roll the *Roddam* far over, almost capsizing it. Then the cloud swept over the ship. The force was so great that the iron chain that held the ship to the buoy snapped. Now adrift, the *Roddam* was bobbing in a wild and turbulent sea. Then a shower of red-hot ash and cinders began to fall, blocking all sunlight and engulfing the *Roddam* in darkness.

Freeman was still in the map room. He could hear the desperate cries of men caught on deck and dying. "They were weird inhuman sounds," he would remember, "like the crying of sea birds in distress."

The hot ash quickly penetrated his hiding place, the intense heat searing his body. "You can imagine what it was like," he later told those who were anxious to hear his tale, "if you think of going to a blacksmith's forge and take up handfuls of red-glowing dust and rub it over your hands and face."

Unable to endure the pain, and knowing he had to get his ship away fast, Freeman ran out on deck.

Men were dying everywhere. Some were on fire and jumping overboard. Others were already dead; their bodies burned beyond recognition.

The entire ship was in flames, ignited by the fall of hot ash. Within minutes, a foot of red incandescent material covered the deck. As Freeman hurried to see who might be saved, his feet slid into pockets of deep ash, burning away his boots. The pain was unbelievable; the suffocation was worse. With each desperate gasp, he gulped down another mass of hot ash, burning this throat and filling his lungs. All the while, he was thinking, "My god, how long is this going to take to kill me?"

After several minutes, the air did begin to clear, and Freeman could again see St. Pierre. The city was a blazing inferno, the framework of stone buildings standing out like dark skeletons.

The volcano was also visible. From the summit spurted a giant fountain, thousands of feet high, of red and orange cinders, the fountain slowly waving side to side. And above the fountain was a colossal tower of black smoke. It was a hellish spectacle that Freeman thought must resemble the Day of Judgment. It was also, as he would later admit, "strangely magnificent."

Unbelievable horrors were all around him. Freeman could see hundreds of people running wildly along the shoreline, flames clinging to them, the poor souls looking "like effigies which had been set alight." Some headed to the sea. Freeman was close enough to hear the bodies sizzle when they entered the water.

Fortunately, the ship's engineer was in the engine room and still alive. Freeman ordered him to start the engines immediately. Freeman then tried to steer the *Roddam* away from shore, but the ship would not turn.

For two hours, he and the few members of his crew still alive struggled to free the ship's rudder of debris washed into the sea by the black cloud. Backward, then forward, they inched the ship until it was free. Then the *Roddam*, a floating furnace, headed for the open sea.

Severely burned and barely able to speak, Freeman took the wheel. His hands were raw and bloody, and so he used his elbows to steer. A crew member stood next to him to wipe blood and sweat from his face.

The *Roddam* passed close to the only other ship still afloat, the *Roraima*. Freeman could see a few people on the *Roraima*, itself a burning wreck, waving frantically at him, but it was in vain for there was nothing he could do. He watched as two men on the *Roraima* jumped overboard and tried to swim for the *Roddam*. He lost sight of them as they were pulled under by a muddy and agitated sea.

Away from St. Pierre, the *Roddam* caught a current and drifted south. After eight hours, just before sundown, the ship entered the harbor at St. Lucia. Chief Surgeon Dennehy and other rescuers boarded the ship and took Freeman and the other survivors to a hospital.

Later, Dennehy returned to the ship. He noticed a curious thing. Sometime soon after the *Roddam* had arrived at St. Pierre, someone had stacked cases of safety matches on deck, readying them for landing. Throughout the ordeal, the matches had failed to light, maintaining a reputation, as stated on each case, "to light only on the box."

———•———

The story of the *Roddam* and the heroic actions of its captain appeared in newspapers two days after the eruption. That was followed by a barrage of newspaper stories that told of other survivors and of the deaths of tens of thousands.

In the United States, people responded with great interest—and with great generosity, contributing to relief funds. President Theodore Roosevelt ordered the Navy to prepare a ship to carry supplies to Martinique. Someone thought to invite a few scientists who might want to go and study what had happened. Surprisingly, not everyone was interested in going.

Nathaniel Shaler of Harvard University, who in an interview for *The New York Times* was called "the country's foremost authority on volcanoes," said that the recent eruption of Mount Pelée had been tragic, but, from the standpoint of science, was of little interest. "A really important eruption," said Shaler who was then dean of science at Harvard, "should be heard fully 2000 miles away." He was, of course, referring to the recent eruption of Krakatoa in Indonesia.

One of the largest volcanic explosions in history, on August 27, 1883, Krakatoa had sent out a series of audible shocks that were heard across southeastern Asia and northern Australia and far out into the Indian Ocean. The explosions also produced giant sea waves that swamped shorelines, drowning thousands of people. In fact, the explosion had been so violent and had thrown so much ash into the air that sunrises and sunsets around the world were turned a deep red for many months. By comparison, as Shaler pointed out, the 1902 explosion of Mount Pelée had been heard no farther than two hundred miles and had not produced any significant sea waves. And ash did not fall farther than a few hundred miles from the volcano.

Furthermore, according to Shaler, the reports coming from the Caribbean were almost certainly exaggerated, a product of an unbridled press anxious to sensationalize any news story. The sudden and complete destruction of an entire city and the deaths of all of its inhabitants could not have been possible from such a small eruption.

The distinguished Harvard professor did admit that one thing might be learned if a scientific team was sent to Martinique. The members of the team could examine the bodies that did exist and determine whether they had died by inhaling poisonous volcanic gas or by being suffocated by volcanic ash. But, other than that, there would be little for them to do. And, as Shaler told the interviewer, he was certainly not interested in going.

Four scientists were found who were willing to join the relief ship, though they, too, on the eve of their departure, expressed concerns about the accuracy of the newspaper accounts and the scientific importance of the eruption. Then a fifth man appeared.

He was Thomas Jaggar, an instructor of geology at Harvard. At first, he, too, had discounted the importance of the eruption, but, then, after reading of the ordeal of Captain Freeman, he decided to go.

A week later, Jaggar was standing among the ruins of St. Pierre, shocked by what he saw. The newspaper reports had been devastatingly correct. The entire city had been swept away. And bodies were everywhere. Nothing like this had ever been reported. And it had happened in just a few minutes.

When he returned to Harvard, he was decided: He would devote his life to a study of volcanoes because, as he told a colleague, "it was a missionary field for in it people were being killed."

And so began fifty years of travel to eruptions in Italy, Alaska, Japan, Central America and the Hawaiian Islands. In 1912 he gave up a comfortable life and secure job in Boston and started a small science station on the rim of Kilauea volcano on the island of Hawaii, drawn to Kilauea because of a rare lake of molten lava. Here he developed the techniques used today to predict eruptions: the collection of volcanic gases, the recording of earthquakes and the measurement of a slight rise or fall of the ground surface as molten lava moves inside a volcano. He also learned how to predict tsunamis and was the first person to warn of an approaching wave. He built the first practical amphibious vehicle and used it to explore volcanic islands, his design later adapted by the United States military during the Second World War for beach landing craft.

All in all, he led a highly accomplished life, though one that is little known today. The obscurity comes from his decision to live the life of a scientific vagabond. It was a decision that ended his marriage and forced him to give up his children. He lost the professional respect of his peers. He lived much of his life in near poverty. At one point, it seemed that he would be forced to abandon his work at Kilauea when he met a woman who changed everything.

She was Isabel Maydwell, a widowed school teacher from California who had come to the Hawaiian Islands to restart her life. And she did—with Jaggar. Together they lived in a small house at the edge of a high cliff that looked toward the lava lake, sharing the work of the small science station started by him, she becoming an astute observer of volcanic activity.

Jaggar's decision to dedicate his life to solving the mystery of volcanoes—why they erupt and how they could be predicted and, possibly, controlled—was inspired by his experiences at St. Pierre. But a desire to confront the great forces of nature came much earlier. It began in the most prosaic of ways with a small boy growing up next to a large river, as far from the fiery parts of the planet as one can imagine.

CHAPTER ONE

THE BISHOP'S SON

R ivers exert an almost mystical hold on the human psyche. Mark Twain used the Mississippi to great effect in *Huckleberry Finn* as a metaphor for free will. The flow of the Jordan River is thought by many to be a pathway to personal salvation. The Rubicon, a small river in northern Italy where Julius Caesar, then a Roman general, crossed with his army, thereby starting a war, is in the lexicon and means to pass a point of no return. Rare, indeed, is the person of a poetic mind who does not see a flowing river as a metaphor for the passage of time or an encounter with rapids as a reminder that life is filled with perils. But to a child a river is exactly what it seems: a place of endless fascination because it is forever changing.

The river in the life of young Thomas Jaggar was the Ohio where it passes Cincinnati. Loons and coots were still common along the riverbanks. Botany and bird life became as important to his education as geography and history. But the lasting lesson of his childhood came from the river and was taught to him by his father, a bishop of the Episcopal Church and head of the Diocese of Southern Ohio, when young Jaggar was thirteen years old.

On February 5, 1884, according to a front-page story in *The New York Times*, "There is much anxiety to-night concerning the condition of the Ohio River." The story continued. "The Ohio River is full from Pittsburg to Cairo and is rising from Cincinnati downward." That night a levee broke at Cincinnati and half of the city was flooded. Fortunately, the Jaggar house was spared, the floodwaters rising to within half-a-city block of its location. But a quarter million people were homeless. The rain continued. After another week the Ohio River was at the highest level ever recorded. And then the situation got worse.

An ice storm unlike any known before or since swept down the Ohio Valley and turned the already swollen river into a raging sea. Waves battered and broke down buildings. The rain changed to sleet, freezing those who were homeless and sleeping outside.

Young Jaggar watched as his father was stirred to action. Bishop Jaggar organized a local relief committee and issued a nationwide plea: "The river towns in the diocese are submerged. A great disaster is upon us. We need help for the present and coming need. Please speak for us."

Today it is all but impossible to realize how unusual—and controversial—such an appeal for humanitarian aid from the entire nation once was, but the United States was then more sectionalized than it is now. Many people felt that communities should provide for themselves, even after disasters. But Bishop Jaggar was a new type of clergyman and he was preaching a new type of doctrine, one that would deeply influence his son and that came to be known as the Social Gospel.

Born in 1839 in New York City, the future bishop liked to tell people he was the perfect New Yorker, Puritan Long Islander on his father's side and Dutch New Yorker on his mother's. He often repeated the story that the Jaggars of Long Island were descended from two brothers who had arrived on the *Mayflower*, a claim that cannot be substantiated because no complete passenger list is known to exist. What is known is that he was a direct descendant of Jeremiah Jaggar, a master of trading ships in the West Indies who arrived in New England sometime before 1646. Jeremiah fought in the Pequot Wars and was one of the founders of Stamford, Connecticut. He was the first of a long line of Jaggars, each one successful in business and conservative in lifestyle, until the arrival of the first Thomas Jaggar.

More impressionable than most young men, the first Thomas Jaggar found his inspiration in the speeches of popular clergy, such as Henry Ward Beecher, who railed his audiences against slavery and other evils. That awakened a deep sense of purpose in the young man who, on the eve of the Civil War, gave up a business career and accepted a religious calling. He was ordained a minister in the family's church, the Protestant Episcopal Church. His first assignment was St. George's on Long Island. Here he met Anna Louisa Lawrence, she also of an old New England family. They married in 1863 and started a family immediately.

Soon a son, Harris King, was born. He lived five years. Two daughters followed, May in 1866 and Anna Louise in 1868. Then another son, Henry Anthony, was born and lived six months. In September 1870, Reverend Jaggar was named rector of the Church of the Holy Trinity in Philadelphia. Four months later, on January 24, 1871, after the family had moved to Philadelphia, the last child was born at home, christened Thomas Augustus Jaggar, Jr., by his father at Holy Trinity.

By then, Reverend Jaggar was embroiled in a test of wills within the Episcopal Church. He was regarded as a low churchman, a derisive term that meant he was less concerned with spiritual matters and sacraments and more concerned with the social work the church could perform, the so-called Social Gospel. And that put him at odds with church officials.

Throughout the 19th century, most clergy in the United States were preaching a doctrine of self-reliance and rugged individualism, two ideals that had served the country well during its western expansion. And many wanted to maintain such ideals. One prominent voice was William Graham Sumner, an Episcopalian minister and a professor of political and social sciences at Yale University. On one occasion, Graham testified in front of a Congressional committee, urging the members not to commit any money—public or private—for humanitarian work, saying such action would impede the natural progress of society and unfairly punished those who had worked hard and had succeeded. But a new type of poverty was rising in the world, one of total hopelessness and despair, due in large part to the rise of industry and the crowding of cities. And people such as Reverend Jaggar tried to counter it. He spoke in favor of hospitals and schools built specifically for the poor. He advocated for child-labor laws and mandatory attendance of children at school.

But many church officials were not in favor of such reformist ideas. And so, in 1874, when a new diocese was made of the southern half of Ohio and it needed a bishop, many doubted that Jaggar would be appointed to the new position. But the people of Ohio campaigned in his favor. A council of bishops met. After a week of deliberations and much politicking—and by a margin of one vote—the council elected Jaggar. He moved his family to Cincinnati immediately where he would run the new diocese.

Within five years, he doubled the number of ministers and increased the number of congregations by half. He organized auxiliaries to raise money to build hospitals and schools. His greatest triumph came in December 1883 when he opened the Hospital of the Protestant Episcopal Church, today the Children's Hospital of Cincinnati, the first such institution in the country devoted exclusively to the treatment of children. It was the Social Gospel at work.

And so it was natural that he acted when floodwaters rose around Cincinnati two months later and he made his plea. And his plea was answered. Tens of thousands of dollars were donated from people across the country. In fact, so much money was received that Bishop Jaggar sent aid to other areas devastated by the flood. Remembering the disaster years later, a friend would write, "The Bishop's high strung and sympathetic nature was intensely stirred. He spared no effort in the work of relief and repair." His work also had a strong influence on his son. Here, as the son could see, was the highest calling—to sacrifice oneself for the betterment of others—an ideal that he would voice years later as "craving service before self."

The Bishop's fame and influence was rising fast. But there was a demon around the corner. In just a few more months, two more tragedies would strike and leave him a lost and listless man.

———•———

On March 27, 1884, just two weeks after the floodwaters had receded and the Ohio River was again flowing at its normal level, the streets of Cincinnati were filled with rioters.

The previous day a jury had decided that William Berner, a seventeen-year-old German immigrant, was guilty of murder and recommended the death sentence. Young Berner had, indeed, been part of a robbery, but not of a murder. The murder had been committed by a different German immigrant. But the people of Cincinnati were in no mood for leniency. Their city was in the midst of a crime spree; nearly a hundred murders had been committed the previous year. And so, when the judge saw the jury's error and commuted young Berner's sentence to twenty years, his action ignited the anger of the people of Cincinnati and they took to the streets.

Known as the Cincinnati Riots, what followed where three days and nights of what is still one of the most destructive riots in American history. More than twenty thousand people fought police who were trying to reestablish order and harassed firefighters who were trying to contain the hundreds of fires set by the rioters. After a second night of violence, the governor of Ohio sent the National Guard to restore order. The soldiers did so with bayonets and Gatling guns, killing scores and wounding hundreds. Stray bullets passed through buildings, killing innocents. Some of those bullets struck the Jaggar house.

After a third night of riots, after the courthouse was burned to the ground and young Berner had been secreted out of the city, Cincinnati was quiet again. Bishop Jaggar then addressed a crowd. He said the recent unrest had been "a concentrated French revolution" and that it had "cleared the political and moral atmosphere." But now was the time for healing and to rebuild.

He reminded the people of Cincinnati that they had been challenged by a flood the previous month and had survived. The city had then been tested by rioters, and had remained intact. He said he was optimistic and that a combination of preaching and generous donations and leadership could undo the destruction. He would probably have been one of those leaders if a third tragedy, this one deeply personal, had not already begun.

During the days and nights of riots, both the bishop's wife and older daughter, May, lay sick and confined to their beds. A doctor was called and examined both patients. He said Mrs. Jaggar was suffering from exhaustion and would recover. But the condition of the daughter was serious. Her illness was unknown.

The mother did recover, but the daughter did not. She continued to weaken. Two months later, on June 2, 1884, with her father sitting by her bedside, May Jaggar, then nineteen years old, "the light of a refined and cultured Christian home," died.

Grieving more than is usual, even after the death of a child, Bishop Jaggar soon became crippled by his own mysterious illness. He went to Philadelphia and consulted Dr. Silas Mitchell, one of the country's leading physicians, a longtime friend and the doctor who had delivered his son.

"If you do not cease work instantly," Mitchell told Jaggar, "you will quit it permanently in less than six months."

And so the Bishop made a decision. He would take a leave-of-absence and, seeking his own recovery, lead his family on a tour of Europe. It would be his son's first great adventure.

The Jaggar family of four—the Bishop, his wife, daughter Louise and son Thomas—sailed from New York on October 3, 1885, and arrived in Liverpool nine days later. En route, the Bishop organized a talent show among the passengers and sold tickets to a performance on the last night at sea. In Liverpool, he donated the cash receipts to the local seaman's orphanage.

In England, his son began a diary. He wrote in it twice a day, summarizing each day's main events and commenting on the family's general health. One of the most frequent entries was "Mama not being well." It would be a common refrain throughout the trip.

Anna Jaggar was, in fact, unwell most of her adult life, though what was the affliction was never determined. She was a melancholy woman. On the back of a large photographic portrait of herself—she is seen sitting stiffly in an upholstered chair, staring, rather severely, into the camera—she penciled the names of her five children, then circled the names of the three dead ones, writing of them, "My babies in heaven." While in Europe, she seldom joined the others, choosing to confine herself in whatever hotel or apartment rooms her husband had secured. Meanwhile, the Bishop, despite his own illness and melancholy feelings, was determined to excite the imaginations of their two surviving children, leading them frequently on daylong trips.

One of the first was to Winchester to see the famous cathedral. Another was to Salisbury to stand among the ruins of Stonehenge. An extended weekend was spent traveling to the Isle of Wight where the father and his two children took a drive to Carisbrooke Castle where, his son noted in his diary, "Charles I was imprisoned."

After two weeks in England, the family crossed the English Channel to Calais where they boarded a train that took them through Paris, then Milan, and finally to Florence, the heart of the Italian Renaissance, where they would reside for the next two months.

The day trips resumed, though young Jaggar reacted with slight enthusiasm. After a day at the Academy of Fine Arts where

Michelangelo's giant statue of David is on display, he recorded little in his diary, except, "a colossal statue with a beautiful expression." Two entire days were spent at the Pitti Palace, once the residence of the Medici family, now the home of endless galleries of fine art, including paintings by Caracciolo, Raphael and Leonardo da Vinci. Then it was two days at the Uffizi Gallery where, among many other masterpieces, Botticelli's *Birth of Venus* is on display. The boy's lack of enthusiasm is easy to understand. He had pent up energy and was being dragged through long corridors filled with unfamiliar art. But his attitude changed after his father took him to the National Central Library in Florence, one of the largest and most important libraries in the world.

Here, in a room filled with stuffed animals and other curiosities of natural history, young Jaggar found a shelf of folios. Among them was a treasure: John James Audubon's *The Birds of America*, all six volumes, and its letterpress companion, *Ornithological Biography*. Each volume was exquisitely bound in beautiful leather. And inside the six volumes were more than four hundred carefully etched, hand-colored prints—each one was two by three feet in size—depicting the private lives, in many cases at true scale, of America's most fascinating birds.

One notable print showed an osprey in flight, holding a fish in its talon. Another was of seven Carolina parrots clutching branches, their bodies positioned at different angles, their heads looking in different directions, the birds seemingly ready to fly out of the page. Three of the parrots have their mouths open and one could imagine hearing them twitter. To see these prints and page through them is to flood the senses. Added to this were the "bird biographies" in which Audubon recounted how he had made the drawings, interweaved with a personal account of the life he enjoyed in the wilderness.

"The margins of the shores of the river at this season amply supplied with game," Audubon wrote when tewnty-two years old and floating on a raft down the Ohio River, passing Cincinnati. It was the 1820s. "We fared well," he wrote. "Whenever we pleased, we landed, struck up a fire and provided as we were with the necessary utensils, procured a good repast."

Here was something familiar to young Jaggar, who had walked the same section of the Ohio River with his father. He also learned that Audubon had lived in Cincinnati for a short time in 1822 while working at a museum stuffing and mounting birds for display and painting backgrounds for exhibits that showed outdoor scenes.

Jaggar returned to the library often to re-examine the books. As years passed and he reflected on what he had found in the library, he came to idolize Audubon, writing that, in his opinion, Audubon had lived the perfect life, that of "a person who could sit motionless for hours watching birds in the wild, sketching and painting them." Jaggar tried to emulate the great ornithologist, but without success. His drawings lacked the careful renderings, his lines the broad sweeps that brought life to Audubon's paintings. Nevertheless, the encounter with Audubon's books helped to set Jaggar on a path toward a study of natural history. The next stop his family would make in Europe would send him even deeper.

After two months in Florence, the Jaggar family traveled again by train, this time south through Rome and on to Naples.

As the train approached the city and its famous bay, Jaggar looked out the window, recording in his diary that "we would pass old ruins; in the distance were snow-clad peaks." As they continued on, "the country grew more rugged," and he saw "peasants dressed in white working the fields and peasant women driving very small donkeys with very big sacks on their backs." Then "all at once we saw Vesuvius," a high conical mountain with "thick masses of blackish yellow sulfurous smokes" rising from its peak.

The train passed along the shoreline of Naples Bay "right under Vesuvius." On the south side of the bay, the Bishop found lodging at a small hotel, securing a room with a view that looked toward Vesuvius. For the son, the choice was excellent. The first night he wrote: "We could see a bright flash come up from the crater every few minutes!"

The timing of their arrival had been fortunate. Just two days earlier, lava had broken out high on the side of the volcano and was running as two streams down the volcano directly in view of their hotel.

A recent winter storm had brought snow to the summit of Vesuvius, and so a trip to see the eruption was delayed. In the meantime, the Bishop took his two children to see the ancient Roman city of Pompeii, buried by an eruption of Vesuvius in 79 A.D.

The exact location of the city had been forgotten until 1748 when the digging of a water well on the side of Vesuvius uncovered several Roman artifacts. More digging uncovered more artifacts. Entire streets and the outlines of buildings were revealed. Eventually, an inscription was found on the side of a stone wall that confirmed the site was, indeed, Pompeii.

Of the many exciting discoveries that have been made at the ancient site, few have been stranger than one made early in 1863 when workers, excavating a Roman house, broke into a small cavity that was filled with human bones, including a skull. The workers removed the bones and studied the inside walls of the cavity. They noticed a pattern of imprints that could have been made only by clothing and hair. There also seemed to be a facial expression preserved. The director of the excavation, Giuseppe Fiorelli, was called to the site.

Fiorelli had the workers pour a mixture of plaster of Paris and glue into the cavity. After it set, he had them scrape away the surrounding material. What was revealed was the cast of a woman exactly as she had been at the moment of death. She was found lying on her back with an arm across her face, as if in an attempt to save her life. Other cavities were soon discovered. They, too, were filled with plaster and glue and a cast was made of each one.

Several of Fiorelli's casts were on display at a small museum at the entrance to Pompeii when the Jaggars visited. In addition to the woman, another showed a man face down with his robe pulled over his head. Another showed a dog in a contorted position pulling on a chain, struggling to save its life. These casts fascinated young Jaggar—as they do visitors today—and he wrote of them in his diary: "They are about as strange and interesting as anything I have ever experienced."

After a day at Pompeii and still unable to make the trip to the summit of Vesuvius because of bad weather, the Bishop took his children on several trips through Naples. They visited a city aquarium where, under the watchful eye of the director, young Jaggar put his hand in a tank and "took a polite shock from an electric ray." Another day they visited the National Museum where he saw "hundreds of bronzes, statues, bas-reliefs, frescoes, mosaics from Pompeii," as well as "Pompeian jewels, money-chests, divans, fish hooks, gladiatorial arms, helmets, and shields and spears."

But, each night, after returning to the hotel, his thoughts returned to the volcano. Once, in an especially poetic mood, he wrote, "The dark blue waters of the Mediterranean stretch away to the right, while on the left looms Vesuvius on whose summit a red flickering light proclaims the presence of unquenchable fires!!!" The three exclamation points were drawn with particular boldness.

———•———

After waiting three weeks, the weather cleared, and the Jaggar family made an ascent of Vesuvius. It was February 25, 1886, "A Red Letter Day," as Jaggar recorded in his diary using, for the first and only time, a red pencil to highlight an entry. His mother also went. It was the only daylong trip she took with the other three.

After breakfast, they rode in a carriage toward Vesuvius. They passed through the village of Portici at the base of the volcano. From there, the carriage took them along a paved road up the side of the volcano. Along the way, whenever the carriage slowed, they were surrounded by "beggars, innumerable boys with flowers, musicians who would walk alongside and not stop playing until given something." At last, after passing through fields of old lava and cinders, close to the summit, the road ended in front of a concrete building that housed a scientific station.

Officially, this was "a meteorological station," as Jaggar recorded in his diary, though, in fact, it was much more. It was a volcano observatory, the first such institution anywhere in the world. The building had been completed in 1848, but, because of revolutions and social strife, it was not occupied by anyone or had any equipment until 1856. Two or three people worked inside of it. And, if Jaggar had inquired of any of them what was known about volcanoes, he would have been disappointed by the answer.

Yes, there was the occasional cataclysmic event—Krakatoa in 1883 and Vesuvius in 79 A.D.—but such explosions were rare. Volcanic activity was not yet seen as a major geologic force. In a popular scientific monograph published in 1881, just five years before the Jaggars' visit to Vesuvius, the writer described eruptions as "entirely mischievous" and the public's perception of their destructive effects as "exaggerated notions." Even at the beginning of the 20th century, a leading geologist would consider volcanic activity to be "local and occasional, not perpetual and worldwide." Jaggar would be one of those who would change this perception.

But, now, as a boy of thirteen, he was anxious to reach the summit and look inside the crater.

Just beyond the observatory building was the famed funicular railway that transported tourists the last thousand feet to the summit. Completed in 1880, it consisted of two large wooden cars, one named "Vesuvio" and the other "Etna." Each car had wooden benches and was capable of carrying up to fifteen people.

The two cars ran on separate wooden rails between lower and upper stations. The cars were connected by a long iron cable that ran around a pulley at the upper station, so that, as one car went up, the other came down. The system was powered by a steam engine at the lower station. Jaggar remembered the ride as "rather rickety and very slow." It took fifteen minutes to make the ascent.

At the upper station, the Jaggars stepped out of the car. They were surrounded immediately by dense fog, which made it impossible to see more than a few hundred feet. Several men offered themselves as guides. The Bishop chose three men, two to carry Mrs. Jaggar in a sedan chair and the other man to lead them to the crater's edge.

"It was a very hard walk," her son wrote of the adventure, "over rough, jagged lumps of warm lava and sulphur, with now and then a bed of sandy ashes into which we sunk up to the ankles."

The air was so heavy with sulfur that he held a handkerchief over his mouth to breath. As he and the others neared the crater's edge, he heard a "dull puffing or priffing sound." He looked into the crater. All he could see was dense fog.

One of the guides said he knew a way down. The Bishop and his children followed the guide, stumbling over a treacherous path, avoiding blocks and rocky overhangs. Finally, on the bottom, more than a hundred feet below the crater rim, they had their first view of red molten lava.

It is hard to convey to someone who has never stood close to where lava is creeping along the ground the strange mixture of senses it invokes. By sight it resembles the slow movement of thick molasses, but with the blinding red glare of an iron foundry. By smell, it has the acridity of the worst sulfur mine. The searing heat can raise welts on bare skin, and so one must keep in constant motion.

The guide showed the Jaggar children how to approach the lava and to use a long wooden stick to retrieve a sample and press copper coins into it. Where the molten stuff was removed, there was always a tremendous hiss, one that Jaggar thought was "like a locomotive on a large scale."

As they began to climb out of the crater, Jaggar stopped and knelt to feel the ground. He stood up quickly, unable to bear the heat for any length of time.

After three more weeks in Naples, with spring finally approaching, the Jaggar family left and headed north, making a final stop in the Italian Alps. Here the son had one final adventure.

One day, he, his father, and his sister had climbed a glacier and were walking across the icy surface. They paused at a high point to admire the view. Just then, they heard a low rumble, "not exactly like thunder," young Jaggar recorded, "but more like Niagara in the distance." A few seconds passed. Then, on one side of a nearby peak, they saw what looked like a large waterfall that gradually dwindled away to a fine stream and then to nothing at all. It had been an avalanche. And it had barely stopped when Jaggar turned and looked at his father and said he hoped to see another one when another happened, bigger than the first. Worried that the icy mass where they were standing might give way, the trio hurried back down the glacier to their lodgings, the son "pleased with our first real alpine experience."

———•———

They had spent an entire year in Europe, though, when they returned to Cincinnati, the Bishop had not recovered from his illness. He wrote a letter to his congregation, ending it with the phrase, "when mind and spirit fail, there is no help." And there would be none for Bishop Jaggar. He offered his resignation. But it was rejected. Instead, the Episcopal Church appointed an assistant who assumed "all duties, powers and authorities" of the diocese. Meanwhile, the Bishop searched for solace.

He found it at a remote spot off the coast of Maine across the Bay of Fundy at Digby Cove in Nova Scotia. Here he purchased several acres, which included a large cabin built of logs. He filled the cabin with books and paintings, most of the latter with religious themes. Here he would remain, his wife and daughter taking an occasional trip, his son returning to preparatory school in Philadelphia.

His son visited during the summer, he and his father taking time to hunt and fish. On a trip in 1888, they shot and stuffed gannet, puffins, ducks and ptarmigan and "caught all the fish we wanted." Another time it was "just father and I and a canvas canoes."

It was the best life the son could imagine. He was living the life of a naturalist, following the lead of Audubon. But it could not last because the father had instilled a deep sense of purpose in his son—a craving of service before self. And to fulfill the purpose the son had to be an educated man.

CHAPTER TWO

YELLOWSTONE

T he modern university with its wide-ranging curricula and liberal ideals was first conceived and instituted by Charles William Eliot, the twenty-fourth president of Harvard University. As soon as he was appointed in 1869, he reduced the student rulebook from forty to five pages and eliminated the requirement that students attend church daily and that they wear black on Sundays. He also removed the privies from Harvard Yard and permitted students to smoke on campus.

As to the curricula, when Eliot arrived, all freshmen took the same courses in Latin, Greek, French and ethics. The second year was filled with physics, chemistry, German and elocution. Before one could graduate, a student had to demonstrate an ability to recite long passages from Gibbon's *History of the Decline and Fall of the Roman Empire* or from the works of one of the writers of the

Enlightenment. Among those that were the most popular among the Harvard faculty were the writings of the Scottish philosopher Dugald Stewart who is little known today but who was once highly praised for his verbal eloquence. Eliot changed this. After years of battling with Harvard's Board of Overseers, he was successful in getting all recitations dropped and in providing students with a list of elective classes, leaving only one that was required to graduate, English composition.

Eliot also convinced the Board of Overseers to expand greatly the types of degrees Harvard granted. No longer were students limited to theology or the classics. They could chose from a wide range of subjects, including, much to the horror of many traditionalists, science or engineering. It was this liberalism, this wide range of choices, which attracted Thomas Jaggar to Harvard, which he entered in 1889. But it was also the liberalism, especially, the loosening of rules, which almost pulled him away.

By the time Jaggar entered Harvard, Eliot had removed the ban on students attending theaters in Cambridge and in Boston. And Jaggar took advantage of this—justifying his actions as "an unusual way to further my education"—by attending performances by some of the great actors and actresses of the age, such as Julia Marlowe, known for her interpretations of Shakespeare, and the ever-popular French actress Sarah Bernhardt. But his real interest was elsewhere. He was fascinated by magicians.

Harry Kellar, who Houdini once proclaimed to be "America's greatest magician," was then working a circuit that included Boston. One night, as Jaggar remembered it, Kellar called him out of the audience and onto the stage. The magician then proceeded to pull a rabbit out of Jaggar's coat and eggs out of his mouth. "Thus I learned the psychology of audiences, how to experiment in public, and how easily deluded is the average mind." The experience inspired him to start his own magic act, which he performed at private parties, advertising on a small cardboard placard: "The sleight of hand of Tom Jaggar, including the Enchanted Hat, Volatile Money and the 'Erratic Kerchief.'" And that led him to act on the stage.

At first, he appeared in crowd scenes, eventually graduating to speaking roles. His first was in *Cymbeline*, one of Shakespeare's least remembered plays. He was the Roman Philario and recited the single line, "My Lord, Posthumus is without." He was thrilled to be on stage and might have made acting his career if a class at Harvard had not attracted his attention.

During his third year, he took a biology class. One day, he picked up a glass slide and slid it under a microscope. He twisted a knurled knob to bring the tiny specimen in focus. Though, years later, in recounting the experience, he failed to mention exactly what he was looking at, he did say that he was amazed that so many tiny animals had so many delicate parts. Suddenly, natural history extended far beyond Audubon's birds. He considered making biology his field of study, but thought it too burdened with laboratory work. He wanted to experience the natural world first hand. He soon got the opportunity.

The next semester he took Geology 4, an introductory class that was popular because it had a reputation for being easy. The professor, Nathaniel Shaler, never took attendance, never gave assignments and required few examinations. In fact, in his first lecture, he advised students not to read textbooks "lest they imbibe wrong notions." The class was so informal that the students called it "Jolly 4" and referred to the professor as "Uncle Nat."

A lecture by Shaler began with hundreds of students seated in a large auditorium waiting for the professor to enter. And when he did, he came into the auditorium through a side door and began to talk immediately. And he talked continuously for an hour. One student, who reminisced about Shaler years later, would say it was the professor's "vivacious and picturesque personality" that carried a lecture. When the professor spoke of the tragedy of soil erosion, "the dust would blow into the eyes of the students." When Shaler characterized a volcano, he became a volcano himself, the students glad "that the ceiling had not gone up along with his arms and voice." But what, specifically, was so special about Shaler? The same student answered. "Well, Shaler aroused enthusiasm." And he certainly did with Thomas Jaggar.

After one particularly dynamic lecture, Jaggar approached the professor and asked if he knew of a project that Jaggar could work on. Shaler was then advocating the draining of swamps along the Atlantic coast to provide more farmland. But to drain and use such land a better understanding was needed of the dynamics of beaches. And so Shaler sent Jaggar to the nearby towns of Nahant and Lynn to study the movement of beach sand.

The young man paced the beaches for weeks, watching the sand form into great scallops whenever the tidewater came in. He then marked and measured the swash marks as the tidewater retreated. Eventually, he settled on a specific topic: He would study the development and movement of ripple marks, the small repetitive dunes that form where water flows over sand. He had watched them form in shallow water along the edge of the Ohio River with his father. Now he had a chance to study them in detail.

"Go ahead," he urged others, "lie on your stomach and watch them." He would kneel down and smooth out the sand ripples with his hand, then watch them reform. He would stand in surf and watch as another wave filled with sand rushed up, cleared suddenly, then retreated, adding yet another thin layer of sand to a growing ripple. For him, the beach became alive, it was "building from the end; it was rippling under wave action."

He rushed to build a large wooden water tank. On the bottom he placed a large glass plate. On the plate he carefully sifted a layer of beach sand. Then, after filling the tank with water, he slowly pushed the plate back and forth. It was nature in reverse, the bed of sand moving instead of the water, but the principle was the same. And, before his eyes, tiny sand ripples formed.

After a year of work, varying grain size, thickness of the sand layer, the depth of the water, and how fast he oscillated the plate, he completed over 130 experiments. After each one, he carefully removed the glass plate—which held one of his "baby beaches"— then made a blueprint of the pattern of regular sand ripples that had formed on the plate. These experiments were the subject of his first scientific paper, "Some conditions of ripple marks," published in the journal *American Geologist*. It was a noble, if forgettable paper, detailing the experimental work, but providing no important scientific insights.

Nevertheless, the experience changed him. No longer was he "an undisciplined student in need of stricter guidance," as he described himself of his early years at Harvard. Instead, he was enthusiastic about learning—and about geology.

He filled notebooks with elaborate drawings that he copied from printed colored plates found in geology textbooks. For a class on economic geology, he transcribed long lists of mining production figures. For one on mineral deposits, he drew cross sections of dozens of mines, noting where the most valuable ores had been found.

His grades were average, but Shaler took note of the budding scholar. And so, in the spring of 1893, when a request came for someone to join a summer expedition to Yellowstone, Shaler sent Thomas Jaggar.

Jaggar wrote to his father that he was "overjoyed at the prospect."

———•———

On June 23, 1893, Harvard University awarded Jaggar a Bachelor of Science degree in geology. Six days later, he boarded a train and headed west for Bozeman, Montana.

Though most of the American West was settled by then, Bozeman was still a wild and raucous place. Founded two years earlier by James J. Hill of the Great Northern Railroad, in 1893 Bozeman had a population of about two thousand and consisted of a few dozen hastily constructed wooden buildings. Several advertised themselves as hotels. Jaggar took lodging in one. He paid the standard rate for a fast-growing western town—a dollar a day for a small private room and a meal. Baths were extra.

Bozeman was where the expedition would begin. And where Jaggar would meet the expedition's leader, Arnold Hague.

Hague knew the geology of the region well. In 1883, as one of five geologists employed by the federal government, he had conducted the first geologic survey of Yellowstone National Park and its surrounding, the park having been established eleven years earlier.

Every summer from 1883 to 1891, Hague had assembled a small team consisting of topographic surveyors and field geologists and sent them to work across the area. In 1892, Congressional funds had been cut for such work. In 1893, Hague had found money to continue the survey, though at a much reduced pace. This year it would be him and one other geologist, a field assistant—Jaggar. For Jaggar, it was ideal because he would be able to work closely with Hague who, according to one long-time associate, was "a gentleman temperate in language and habits at all times—even with mules." Hague was also known to take a particular interest "in the work of a beginner."

When Jaggar met Hague, the expedition's leader was busy hiring men who would do the routine work of tending the pack animals, hunting, transporting the camp equipment and maintaining camp. That day, Hague hired two men who had worked for him before. Both would again prove to be reliable men.

Next Hague interviewed men who wanted to be the expedition's cook. The first man to apply had also once worked for Hague, but, according to Jaggar, Hague thought the man was "too consumptive" and refused to hire him. Instead, Hague spent the remainder of the day testing new men. He finally settled on one named John Anderson, a former slave who claimed to have been a soldier in General Custer's Big Horn expedition. The first night, Anderson prepared a meal that included fresh-baked bread and strawberry shortcake. It was a gratifying meal, but, after two months in the mountains, Jaggar would complain that Anderson's cooking had become "too economical."

Jaggar's first assignment was to find a horse. He chose a stallion. When he tried to get on and take his first ride, the stallion bucked so hard that Jaggar fell head first onto a folded tent. Hague now intervened and found a horse more suited to the Harvard student, a bay mare named Bessy, which Jaggar later admitted was "slow and harmless." For the first few days of travel, Jaggar rode sideways with one leg hung over the pommel until his body adjusted "to the vagaries of the saddle."

Hague put his field assistant in charge of compiling an inventory of the expedition's equipment. Tents, saddles, spurs, blankets and an assortment of field instruments were set out. The list included thermometers, a surveyor's level and a compass. In a small personal bag, Jaggar stored soap and a flask of brandy. He was making good headway packing the expedition's equipment, delayed only once when "a small boy came around and asked twelve thousand seven hundred and thirty questions."

The expedition departed Bozeman on July 15. The five men rode east, covering fifty miles the first day, following the route that I-90 takes today. The first night they camped near the opening to Boulder Canyon. Along the way, Jaggar noted there was a log cabin every few miles, most with a sign that said "Saloon."

Turning south, the men left forested land and entered barren, rolling country. Here Jaggar saw thousands of prairie dogs "and their owl boarders." By the third day, the expedition was winding its way slowly through a glacial moraine piled high with boulders. That night Jaggar heard his first coyote.

It was another week of travel before the expedition reached the area where Hague planned to begin the geologic work. The objective was the Absaroka Range that forms the mountain barrier east of Yellowstone National Park. This mountain mass covers two thousand square miles, almost the size of the park, and was one of the last areas of the American West to be explored, bypassed by earlier explorers and mountain men because of the rugged terrain. Those few who had entered the region reported that only two trails ran through it and that those were covered by snow most of the year, which made them impassable except during a few weeks during summer months.

Once in the Absaroka Mountains, the expedition advanced only a few miles a day. A new camp was made almost every night. Here were elk, grouse, blackmail deer, antelope, rattlesnakes, prairie dogs, whistling martens, owls and badgers. Jaggar learned to live with kicking mules. He recorded seeing "colorful canyons and towering stocks of old volcanoes." Within some of the rock layers, he found fragments of petrified trees and imprints of fossil leaves. Most of the surrounding peaks rose higher than 10,000 feet; some were higher than 13,000 feet. He scaled several of them. Almost everywhere he looked he saw millions of years of earth history exposed in the sides of mountains. In comparison, he wrote, "Our schoolbook history is pretty small."

A typical day began when Hague and Jaggar climbed a nearby peak. From there, they would study the rock layers that were visible for miles. For an hour or more, Hague would sit and gaze through his field glasses—which he was always losing, sending Jaggar back to recover them. When he was satisfied with what he had seen, Hague would turn to his field assistant and tell him what questions they would address that day. Was a line of hills the former root that once fed lava to a now dead volcano? Did a nearby valley contain rocks of an ancient seabed that might be filled with fossils?

Each day passed with the two of them riding together, far ahead of the others. "We rarely exchanged a word," Jaggar noted, "using every minute to observe the details of the rocks above and on the ground, topography and scenery in general." Occasionally, they stopped and Hague asked Jaggar to dismount and take a hammer and knock out a rock sample or use a barometer or a compass and take a reading. At times, Jaggar rode ahead, other times, Hague did. "The advantage of this trip as a geological education cannot be overestimated," Jaggar wrote one night in camp midway through the expedition. "The dike formations of these canyons are as fine for study as any that can be imagined."

In addition to being Hague's field assistant, Jaggar also served as the expedition's photographer. In the evening, for a darkroom, he dug a shallow trench in a nearby stream. He set poles over it and covered the poles with horse blankets. He then crawled inside, carrying bottles filled with chemicals and the glass plates he had exposed that day. After the plates were developed, he placed them in the trench to be rinsed thoroughly all night by flowing water.

As a bonus, after a day's work, Jaggar was free to fish and hunt. Once, about two weeks after leaving Bozeman, he "fished afternoon and evening," catching eighteen fish, averaging half-a-pound each. Another time, he hunted "a great flock of grouse" and "shot two with rifle."

After one day of geologizing, the men were camped at the base of a cliff when Anderson, the camp cook, called out, "Mr. Jaggar, I smell sheep up on that shelf." Anderson leaned his rifle against a steep slope of loose rocks, muzzle upward. He then proceeded to climb up to get a better look, knocking down rocks as he went. Unable to shake a geologic instinct, Jaggar began to examine the rocks as they tumbled down toward him. He saw that they were a Cambrian limestone that contained trilobites, an extinct animal that once crawled on the floor of an ancient sea. As he began to sort through the rocks, looking for the best specimens, he noticed, too late, that some rocks were falling and hitting the rifle, causing it to slide. As he later told the story:

> I grabbed for the muzzle pointed toward my throat, the stock wiggling right and left. The gun went off and I felt a nick in my ankle. The cook had left a cartridge in the barrel with the hammer resting on it, but the nick was made by a pebble ploughed up by the bullet. So the trilobites took a shot at me!

During the last two weeks of the expedition, Hague led the party through Yellowstone National Park. Here Jaggar rode through geyser fields and saw colorful mineral-encrusted streams. It was among such fantastic sights, surrounded by the ever-present grandeur of mountains, that Hague became introspective. Often, when they were camped, Jaggar would watch Hague walk off by himself to look at a waterfall or to peer into a canyon. And, whenever he did, Jaggar noticed that Hague always removed his hat. Afterward, the expedition's leader would return and take his assistant aside and direct his attention not only to the natural wonders all around them, but tell him of the intangibles, such as the instinctive movement of elk through a forest or the grandeur of a winter storm. Hague said it was such things that had sustained him during a lifetime of living on a frontier. And he had stories to tell.

Born in Boston in 1840, Hague had attended Yale University where he had been influenced by two popular books, Darwin's *Voyage of the Beagle* and Humbolt's *Cosmos*. They had infused him with a sense of wanderlust. And so, when given the chance, he joined one of the geologic parties then surveying the American West.

The party he joined was the Fortieth Parallel Exploring Expedition led by Clarence King, himself a notable character in the history of American science and culture. King was the friend of presidents and a member of a tight little social group that called itself the Five of Hearts that met in a three-story mansion on Lafayette Square in Washington, D.C., across from the White House.

Hague joined King's expedition in the winter of 1867, spending his first months at the Comstock Lode near Virginia City, Nevada, the site of the richest gold and silver deposits on the continent. Though he began by exploring the mines and describing the surrounding geology, he soon turned to improving what was known as the Washoe process, a method that was used to increase the recovery of gold and silver from material discarded by the mines. In that, he succeeded, showing miners how to increase their yields by half, earning himself the gratitude of the owners of the Comstock Lode. For Hague, who was indifferent to riches—his father was a Baptist minister who extolled the virtue of hard work over the sin of greed—it was enough to solve a practical problem. For Jaggar, it was an example of a problem that could only be solved by someone who was living on a frontier.

After working for five years on a geologic survey of the American West, Hague and King sailed for the Hawaiian Islands, staying in Honolulu only long enough to catch the next ship for Hilo on the island of Hawaii. Once there, they hired horses and spent an entire day riding a difficult trail, mostly through dense forest, to the summit of Kilauea volcano. As Hague described the adventure, he recounted how at the summit they found "a molten lake ringed by fire" that occasionally sent streams of lava in the direction of their campsite. Once, to Hague's obvious delight, a jet of gas and molten rock shot into the air right in front of him, a whirling tower of red incandescent rock that "licked across the surface" of the lake.

To Jaggar, this was live geology. He could only imagine what it must feel like to stand near to such excitement.

The story of travel to the Hawaiian islands and of standing at the top of an erupting volcano, as well as Hague's later adventures as a government geologist in Guatemala studying that country's mines and volcanic districts, as a mine assayer in China where he was once caught in a dust storm so thick that it completely blocked out the sun, and as a mountaineer who climbed several prominent peaks in the northwestern United States, including the volcano Mount Hood, stirred Jaggar's imagination all the more. Which may be why, when the expedition completed its work and returned to Bozeman, Jaggar wrote that he was disappointed "to come out of the back wood to this tourist place."

———•———

In the official report of the 1893 expedition, Hague recorded that he had been accompanied by "Mr. T.A. Jaggar, jr.," who "came highly recommended" and "rendered efficient aid." In recognition of his assistant's work, Hague named a peak in the Absaroka Range "Jaggar Peak." A stream leading north of the mountain is known today as Jaggar Creek.

Jaggar returned to Harvard and took a fifth year of classes, earning a master's degree in geology. Instead of returning to the American West, he decided to spend the next year in Europe studying at German universities, well aware that Hague had done the same after his graduation from Yale.

With introductory letters from his Harvard professors, Jaggar enrolled for the 1894 fall semester at the University of Munich.

Here the main attractions were a magnificent mineral collection and a forceful lecturer, Karl Alfred von Zittel.

An authority on fossils, von Zittel was the embodiment of German pride. Before each lecture, the professor's assistant came into the room and arranged diagrams on a rack. Then the students stood and, as Jaggar remembered it, "his majesty entered."

Von Zittel grabbed a long rattan pointer and commanded the students' attention, calling for recitations. He walked around the room whacking the diagrams for emphasis, all the while making allusions to the inadequacies of American science and its scientists.

There were other Americans in the class. One was Charles Palache, a recent graduate of the University of California. On weekends, Jaggar and Palache traveled together across the German countryside looking for quarries where they might collect rocks. Jaggar carried a miner's hammer that he had named "Umslopagass," a reference to the Zulu chief in Rider Haggard's *Allen Quatermain*, a recently published novel with suggestive sex scenes. Whenever either of the young men encountered a rock outcrop that would not easily yield a sample, he would call out to the other, "Umslopagass, come quick!"

Jaggar and Palache spent the next semester at the university in Heidelberg where the star professor was Karl Rosenbusch, a man who could chat endlessly with students while dropping cigar ashes on their clothing. On the first day of class, Rosenbusch assigned each student a single mineral to examine the entire semester. Palache delved into the work, fascinated by every new aspect and nuance he could discover. For Jaggar, however, the work was tedious. He wanted things that were "moving, changing, evolving." He had hoped to have access to a high-temperature oven that he could use to melt the sample, then watch it reform. Instead, he spent two months peering through a microscope, trying to imagine what every minute feature might mean.

Fortunately, as his year in Germany was ending, a telegram arrived from Cambridge asking Jaggar if he was interested in returning to Harvard as an instructor of geology. He jumped at the chance.

He and Palache sailed back to the United States with Jaggar returning to Harvard and Palache to California. Of the voyage, Palache would recall "little of the crossing save that Jaggar proved a poor sailor and that we met a jolly party of American girls." Their time in Germany had cemented a friendship. When Palache was unable to find a position at the University of California, Jaggar recommended him to the curator of Harvard's mineralogical collection who hired Palache as an assistant. For the next year, the Californian studied and arranged the collection by day and, armed with a rifle, guarded it by night.

Meanwhile, Jaggar began teaching classes at Harvard. The first semester he taught one about the geology near Boston, taking students on walking trips to see outcrops and to collect fossils. The second semester the Harvard faculty allowed him to develop a new course, one of his own design. "Experimental and Dynamical Geology" was a series of simple experiments that demonstrated geologic processes in action. The first time he offered the course, one student enrolled. During the next ten years he improved the experiments, though the enrollment never exceeded eleven students.

Besides showing how ripple marks could be reproduced in a water tank, Jaggar built another large wooden box and filled this one with layers of different materials—sand, marble dust, coal dust. He then set the box at an angle and sprayed water over it to show how the Grand Canyon may have formed. He filled yet another box with alternating layers of plaster of Paris and coal dust, then squeezed beeswax between layers to show how magma might intrude into the earth's crust and cause the surface to dome upward. Occasionally, the beeswax broke through the topmost layer and a tiny volcano formed.

His grandest and most popular demonstration was that of a geyser. Jaggar assembled a maze of water-filled flasks connected by glass tubes, all the while telling students of his own experiences at Yellowstone. He lit a burner beneath one of the flasks. Within minutes, water shot up from a vertical pipette at one end of the tubing, the spouting water rising as high as four feet, pulses coming at regular intervals of ninety seconds.

One spring, he took his portable geyser on a tour of local museums and colleges. A reporter for the *Washington Post* saw one demonstration and was so impressed that he pronounced Jaggar "a new type of geologist" one who could show "in miniature, by means of specially contrived apparatus, the great geologic processes of nature." The reporter ended his praise of Jaggar and his performance as being "like that of a magician."

While continuing to work as a Harvard instructor, Jaggar also pursued a doctorate degree. To fulfill the degree requirements, he submitted the invention of a machine that measured the hardness of minerals while pulling a diamond-tipped needle across a small mineral sample. It was an ingenious machine, yet, some members of the Harvard faculty questioned whether the invention of "a mere instrument" was sufficient for Harvard's highest degree. And so Jaggar worked another year and added a second part, a description of a rock outcrop near Boston. Years later, when reflecting on the additional requirement, he would write, it was "university teachers, in my own experiences at Munich, Heidelberg and Harvard, [that] defeated their own sciences for me as a budding scientist."

Someday, as he was coming to realize, he would have to choose between two worlds, "between museums and field, between the easy thing of collections, fine microscopes and scientific societies, and the hard thing of exploring the globe."

But, for now, the decision could be delayed because Hague had invited him to work again in Yellowstone.

———•———

On July 5, 1897, he was back in Bozeman, Montana. This time, Hague asked him to purchase the horses they would need for the expedition. It was "a nasty business," he would recall, to "have to deal with a wild class of men and they all lie." In the end, after an entire day of intense dealing, he bought six riding horses and seven pack mules, two of the latter ones wild, knowing, from his experience four years earlier, that "Mr. Hague was strongly in favor of using mules."

The two men worked again in the Absaroka Mountains, then Yellowstone National Park, this time, Hague treating him more like a colleague than an assistant. They enjoyed late nights around a campfire, exchanging opinions about politics and religion and discussing such lofty topics as the philosophy of happiness. Hague also encouraged Jaggar to explore the terrain on his own.

One day, while in the national park, Jaggar was directed toward a forbidding place known as Death Gulch. It began with a long climb up a steep slope and through thick timber. At the end of the timber was the entrance to a barren ravine, its walls stained white and yellow with salts leached from the ground. Jaggar labored up the ravine, his breathing heavy because the air was filled the smell of sulfur, though, when he later wrote of the experience, he introduced his adventure in a light-hearted, if slightly vulgar manner.

> Cases of asphyxiation by gas have been very frequently reported of late years, and we commonly associate with such reports the idea of a second-rate hotel and an unsophisticated gentleman who blows out the gas. Such incidents we connect with the supercivilisation of the nineteenth century, but it is none the less true that Nature furnishes similar accidents, and that in regions far remote from the haunts of men.

The ravine was about fifty feet deep and a hundred or so feet wide. Death Gulch had a stream of cold clear water running down its center. Jaggar bent to taste the water. It was soured by sulfur.

He continued to climb through this "frightfully weird and dismal place," finally stopping when, a short distance ahead, he saw the latest victims.

They were a group "of huge recumbent bears." The one nearest to him was lying with its nose between its paws, looking exactly like a huge dog asleep. Jaggar approached cautiously. He threw a pebble and struck it on its side. The only response was a belch of noxious gas that almost overwhelmed him.

He examined the carcass and decided it was a young grizzly. Drops of thick, dark-red blood stained its nostrils and the ground beneath. Five other bears lay nearby in various stages of decomposition. One grizzly had died so recently that its tracks were still visible and could have been followed back in the direction it had come.

The deaths had been caused by carbon dioxide, a gas common in volcanic regions and emitted from the ground. Heavier than air, the invisible and odorless gas had accumulated in the lower reaches of the ravine. Each bear had succumbed to the gas.

Fortunately, a strong wind was blowing and bringing fresh air into the ravine the day Jaggar visited. Even so, he lit a few matches and threw them into small hollows and cavities. Each one burned brightly, indicating the air was safe to breath. The smell of sulfur, common in volcanic areas, was plentiful, Jaggar feeling a burning sensation deep in his lungs long after he left the area. It was a small discomfort, so he thought, for such a rare adventure.

———•———

Jaggar returned to the American West every summer for the next few years. At first, his fieldwork was focused on the volcanic rocks near the Black Hills of South Dakota and around Devil's Tower in Wyoming. Later, with Palache, he mapped the geology of the Bradshaw Mountains in Arizona, showing that the area had once been buried by lava flows, then eroded and, finally, uplifted.

In 1899, after his summer work ended and just before he returned to Harvard, he wrote a letter to his father. "I have received your two letters, and send you herewith the $100 I promised." He continued. "I hope you will come on at Xmas, for there are several things to be talked over, and some of them I will tell you about now." The first was the possibility that he would be leaving the United States.

The Royal British University in Melbourne, Australia, was looking for someone to teach geology. And Jaggar was one of three finalists. He wrote to his father that he was anxious for the job, seeing it was "a great opportunity for travel and for widening my experiences." Unlike Germany, Australia would be an adventure, more in line with the American West.

And Jaggar had other news. "If I should get the Australian position I should probably marry. The woman in question is entirely new to you, and so new to me that I don't quite know where I am at." She was Charlotte Gage, trained in Europe as an opera singer and "a descendent of the old tory governor." Jaggar told his father that Miss Gage "will have seventy or eighty thousand dollars of her own when her mother dies." And that she was divorced.

"Now on general principles," he continued, in writing to his father, "the idea of marrying a divorcee is one that I don't like any better than you do, but unfortunately, general principles do not stand by us when we get into the real life of the exception." Then, to reassure his father, "She is womanly and pure, not a grain of stage-struck fly-away about her." She was "a brilliant social figure while thoroughly accomplished in housewifely pursuits." He ended the letter by admitting "I have flirted so much that I don't possess the power yet of taking myself seriously."

In the end, he was not hired for the job in Australia nor did he marry Miss Gage. Instead, years later, when he recalled the events of 1899, he would remember two other events that "affected the rest of my life."

The first happened in September when a swarm of intense earthquakes were felt around Yakutat Bay in Alaska. Reading newspaper accounts of the faraway activity, Jaggar was impressed by eight prospectors who, with no scientific training, had devised a simple way to count earthquakes.

The prospectors had hung two hunting knives from strings, suspended so that the points almost touched. Whenever the ground shook, even slightly, the knives jiggled. With this crude device and a pocket watch, during the next week, the prospectors timed fifty-two earthquakes. Then, at 9:30 A.M. on September 10, their makeshift sensor, as well as their campsite, was destroyed when, as one prospector remembered, "There came another severe shock that was enough to throw a man off his feet."

The eight men were sitting inside a large tent when the strong earthquake struck. As soon as the shaking began, those who could get to their feet raced out, one man momentarily delayed when he was thrown onto the camp stove. Two others kept in a corner, holding tightly to a tent pole. The shaking lasted almost three minutes, one man remembering the ground "cutting some of the queerest capers imaginable."

The ground shaking was so strong that it caused a wall of water to rise in Yakutat Bay and run up onto shore. It washed over the eight prospectors, sweeping them over a forty-foot-high hill and landing them atop the crest of ridge. Gaining their feet, they ran along the crest, as one prospector remembered it, "the [water] boiling and seething at [their] feet." All eight survived. It took them four days to reach the town of Yakutat, which, though the wall of water had not reached the town, it was deserted. After some searching, the prospectors found the citizens of Yakutat camped on a nearby hill that, to this day, is still known as "Shivering Hill."

According to Jaggar, "The Yakutat earthquake snapped on an astonished world, though most of the world didn't known it." Actually, the earthquake "snapped" mostly on him, awakening him to how much there was to learn about the earth and its forces. And, as he knew, the Yakutat earthquake and others like it were happening and causing havoc and going unstudied and, in some cases, unreported elsewhere in the world.

The second event of 1899 to affect the remainder of Jaggar's life was a request he received from Charles Doolittle Walcott, Director of the United States Geological Survey. The Hawaiian Islands had been annexed the previous year and, knowing of Jaggar's interest in studying volcanic rocks, Walcott asked the Harvard instructor to put together a plan for a geological survey of the islands.

Jaggar thought back to his field experiences with Hague and listed the requirements. He would need two field assistants, two packers, a cook, a half-dozen mules, a reliable set of surveying equipment and enough money to support the work for three years, the length of time he thought was needed to survey all of the islands. Jaggar submitted the plan to Walcott in January 1901, then waited for a response.

During the next year, he was elected to the American Academy of Arts and Sciences. He continued to teach at Harvard, receiving an outstanding teaching award. And he worked the next summer in Arizona with Palache. His future seemed secure. And, yet, recalling this time, he would write, "I dislike geology." What had gone wrong for this promising Harvard graduate?

His dislike of geology grew from "its concentration on mining interests," which meant one had to work in "secrecy and its devotion to profits." It was completely opposed to the ideal of the Social Gospel.

In January 1902, Jaggar wrote to Walcott, reminding him that a year had passed since he had submitted a plan for a geologic survey of the Hawaiian Islands. In the letter, Jaggar volunteered that "if I were to go to Hawaii, I should resign from the University." Walcott responded. "Your plan of organization for geologic work in the Hawaiian Islands is duly received and considered. It is placed on file for future consideration whenever Congress shall take such action in regard to an appropriation for the proposed work."

The response was a disappointment. An opportunity had slipped away. And no others were in sight.

But, four months later, a natural disaster would stun the world and change his life. It was all the more remarkable because no one knew that such a disaster was even possible.

CHAPTER THREE

THE CARIBBEAN

T he dozen islands of the Lesser Antilles, among the smallest islands of the Caribbean, lie along a great arc that begins east of Puerto Rico, swings southward into the Atlantic, then curls back, ending at Grenada, just north of South America. Lying along this arc are the islands of St. Kitts, Guadeloupe, Dominica, Martinique, St. Lucia and St. Vincent. Each name invokes an image of an island lush with tropical vegetation and surrounded by a serene blue sea. But looks are deceiving because each island is a potential Vesuvius in eruption because each island was born of volcanic fire.

The two most active volcanoes of the Lesser Antilles share the French name *Soufrière*, which means "sulfur fire." One of the Soufrière volcanoes forms the western half of Guadeloupe and the other most of the island of St. Vincent. Midway between them is the island of Martinique, a French colony, fifty miles in its longest extent and no more than twenty miles wide. And it has its own volcano, Mount Pelée.

Mount Pelée dominates the northern half of Martinique. Its summit rises higher than four thousand feet and its steep slopes are drained by a dozen rivers that have sulfated the volcano into a series of pie-shaped segments that diverge from the summit and run to the coast. Most of the coastline is lined by high sea cliffs, giving the appearance that the edge of the volcano was carved by a giant trimming knife. In only one place is there a narrow coastal plain. It is on this plain that the people of Martinique built their largest and most vibrant city—St. Pierre.

Known to 19th-century travelers as "the Pearl of the Caribbean," St. Pierre was an essential stop for anyone who sailed the Caribbean in that era. The pattern of its streets and much of its architecture was reminiscent of an old French town, though the people of St. Pierre enjoyed a much better climate. Buildings were made of stone and consisted of only one or two stories. Black iron-wrought railings outlined balconies. Large windows allowed sea breezes to blow through. In case of a hurricane, strong wooden shutters could be closed, protecting the inhabitants.

The city had two main streets that ran parallel to the shoreline. Each one had a streetcar line that was run by female conductors. There was an electrical power plant and a telephone system that connected St. Pierre to every village on the island. The city had two banks.

Four newspapers were printed. An opera house offered stage productions from Paris. But there was a dark side.

Alleys lined with rum shops and brothels ran from the main streets down to the waterfront. Local law prohibited such establishments from having outside signs, and so patrons had to step inside to see what was offered. And it was of the vilest kind. Rooms were dank and dark. Robberies were common. Some women serviced more than a dozen customers a night. It was such wickedness and permissiveness that caused some to say that St. Pierre would be struck soon by a vengeful God. And that prophecy seemed that it might come true when, in early 1902, the people of St. Pierre first reported the smell of hellish sulfur wafting down from the summit of their volcano.

No one would ever remember exactly when the disagreeable odor was first detected, except that it had been during a morning and that, by afternoon, a sea breeze had carried it away.

Then, sometime in late March—again, no one could remember when—an unusual cloud, light gray in color, shot up from the summit of Mount Pelée. For a moment afterwards, the ground shook. Those who were long-time inhabitants said it was nothing to bother about. A similar thing had happened more than fifty years earlier when, for about a month, the volcano had sent up a series of small puffs of gray ash that settled quietly over the city. Yes, the old-timers remembered, the ground had shaken occasionally then, too. What was happening now, in 1902, was just a replay of the earlier event. And there was no reason for concern. The puffs and shakings were entertainment and should be enjoyed. But not everyone was convinced. Some people quietly hustled off to church to confess their sins.

After a day or so of puffs of ash and brief earthquakes, weeks of calm followed. That was broken on April 23 when three quick earthquakes rocked the island. The wife of the American consul on Martinique, Clara Prentiss, who lived in St. Pierre, wrote to her sister about the events.

At first, Prentiss thought someone was at the door, but, when a second, then a third shaking was felt, the last one strong enough to rattle dishes, she ran to the window to see what was happening. A column of ash, darker than the earlier ones, was rising high above the volcano. Minutes later, a rain of fine ash fell over the city. That afternoon, to calm any unease the people of St. Pierre might feel about the volcano, the island's colonial governor, Louis Mouttet, who had lived on Martinique for nine months, issued a statement saying there was no danger from the volcano because, if Mount Pelée did erupt, the city of St. Pierre and its inhabitants would be protected by a high ridge that stood between the city and the volcano, a ridge that had protected St. Pierre many times from the high winds of hurricanes. In the same way, the statement concluded, St. Pierre would be safe from the volcano.

After a day of quiet, Mount Pelée exploded again. This time it was heard across the entire island. At first, most people thought it was just the firing of distant cannon until they saw another plume of ash, darker than any so far, rising from the volcano. The eruption continued for several hours. And, again, powdery material fell over St. Pierre, enough to silence the familiar clicks of carriage wheels rolling over the city streets.

On May 3, Clara Prentiss wrote another letter to her sister, keeping her informed of what was happening in St. Pierre. She and her husband were worried. They had made plans to leave the city quickly if the volcano seemed ready to burst. A schooner, the *R.J. Morse*, owned by a wealthy American, had arrived and, if the situation seemed serious, Clara and her husband would grab their most valuable possessions and race to the waterfront and sail away.

The next night the volcano exploded again, this time more violent than before. It rained heavily that night, the rain mixing with volcanic ash, forming muddy slurries that raced down streams and rivers. The Blanche River, three miles north of St. Pierre, usually clear, was a raging flood of mud that buried a sugar mill, killing dozens of workers. When the flood reached the sea, it sent out a huge wave that swept around the island, swamping low wooden piers at St. Pierre, carrying away some cargo. The sea quickly calmed, but not the people.

The fatalities at the sugar mill and the destruction by the sea wave raised so much concern that Governor Mouttet issued a second statement. "There is nothing in the activity of Mount Pelee that warrants departure from St. Pierre." He added, "The relative positions of craters and valleys opening toward the sea supports the conclusion that the safety of St. Pierre is assured." Then, as a demonstration of his faith in his own pronouncement, Mouttet and his family and several key government officials, who all resided in

Fort-de-France on the south side of the island, moved to St. Pierre where they would stay, Mouttet said, until an election was completed five days later.

On May 7, three days after his second statement, Mouttet issued another one. This time he announced that the volcano of Soufrière on the island of St. Vincent, a hundred miles to the south, had exploded, killing thousands of people. At first, the people of St. Pierre were worried by the news, then became quietly relieved when a rumor spread that the volcano on St. Vincent and their volcano were connected. The explosion of Soufrière, it was reasoned, had released pressure beneath Mount Pelée, making their volcano less threatening.

That night, another storm hit the island. By morning, May 8, the weather was clear. Just before daybreak, the sugar-carrier *Roraima* from New York entered the harbor with several passengers on board. An hour later, the British steamer *Roddam* arrived. Its captain, Edward Freeman, anchored his ship at the south end of St. Pierre. Thirty minutes later, about 7:30 A.M., while waiting for the ship's agent to arrive, Freeman watched as a small boat filled with tourists from Fort-de-France arrived at St. Pierre. They planned to spend the day in the city, hoping to see some volcanic activity.

That morning, the telephone operator at Fort-de-France, twenty miles south of St. Pierre, was making his usual morning calls, checking the wires and chatting with friends. At a few minutes before 8 A.M., he was talking to someone in St. Pierre who said that a dark fog was descending on the city. The person in St. Pierre said he was going to take a look and would call back.

Minutes passed and the operator in Fort-de-France heard clattering on the roof. He went outside and saw that a hail of stones was falling from the sky. He hurried back and called his friend in St. Pierre. The friend answered, barely able to speak. The operator shouted into the headset, and then listened for a response. He heard a cry of intense pain followed, as he told the story, "by an incredible sound, like that of an enormous block of iron falling on a metal roof."

The operator hung on. Seconds passed. He pressed the headset against his ear hoping to hear more when a surge of electricity, racing through the wires, sent a violent shock through his head.

———•———

Thursday, May 8, 1902, was a typical dreary mid-spring day in Massachusetts. The sky had been overcast for a week. That morning a cold rain fell in Cambridge and in Boston.

For Jaggar, the day had been an ordinary one. During the afternoon, he taught two classes, "Experimental Geology" and "Advanced Geological Field Work." Some time during the day, he had probably found time to add to an evening public lecture he was planning to give the following week, a lecture entitled "Volcanoes and You." On Saturday, he was planning to lead a group of Harvard geology students on a hike through the nearby Waschsett Mountains to show them evidence of recent glaciation. During a meeting held earlier that week, he had told everyone who planned to join him on the hike that they would be leaving the Harvard campus promptly at 8 A.M. and would be returning after dark. The trip would be made regardless of the weather.

That evening, May 8, as he was returning home, news of the eruption of Mount Pelée and the destruction of St. Pierre was racing through telegraph wires. He learned of the eruption the next day.

The headline in *The Boston Globe* read: "Killed 25,000! City of St. Pierre Wiped Out. Mount Pelee in Eruption." Beneath it was a short article, barely 200 words that offered few details. With so little information, it is not surprising that news of the eruption made no immediate impression on Jaggar.

The next day, Saturday, the entire front page and the top half of the second page of the *Globe* were filled with news articles about the destruction of St. Pierre and background stories about Martinique. Included on the front page was an account of how Captain Freeman of Suffolk, England, had managed to save his ship, the *Roddam*, from the disaster. That, though, was enough for Jaggar to act. He went looking for Harvard President Eliot.

He found Eliot on campus. He reminded the president that they had spoken before about sending Jaggar to study a major eruption. But Eliot advised patience. There was no way to know if the newspaper reports were accurate. They should wait and see if the reports coming from the Caribbean were confirmed.

By Sunday, May 11, three days after the eruption, much of the world's attention was focused on Martinique. In the United States, churches and private charities were collecting money to send to survivors. That afternoon, in Washington, D.C., the Senate held a special session and unanimously approved $100,000 for relief supplies for Martinique. The bill was sent to the House of Representatives that evening. Congressman Oscar Underwood of Alabama, a Democrat who was thinking of challenging Republican President Theodore Roosevelt in the next presidential election, advised his House colleagues against haste. The Senate had gone through "a legislative spasm," according to Underwood. He suggested that a special committee be formed to assess the situation and to write an official report before the House took action.

The next day, Monday, after conferring with the French ambassador at the White House, President Roosevelt sent a request to Congress for $500,000 in aid for Martinique. Leaders of the House and the Senate met and, after deciding to ignore Underwood's suggestion, agreed on $200,000. The bill passed both houses of Congress and was rushed to the President, who signed it that night.

The same day, Jaggar sought out Eliot again. This time Eliot said he had talked with Secretary of the Navy William Moody, a Harvard man, Class of 1876, who said a Navy ship was being prepared to carry relief supplies to Martinique. If Jaggar could get to New York before the ship sailed, he would be taken to Martinique.

The ship was the USS *Dixie*, which had just returned from a year-long cruise to the Mediterranean and the Middle East. Originally built as a private steamship, the *Dixie* had been purchased by the Navy in 1898 at the beginning of the Spanish-American War to transport supplies to Cuba and Puerto Rico because it had more cargo capacity than a regular Navy ship.

Not to be left out of what had quickly developed into a major news story, also on Monday, May 12, Commanding General Nelson Miles of the United States Army ordered the commissary to purchase 800,000 rations for Martinique, enough to feed 40,000 people for twenty days. The rations would be carried on the *Dixie*. More than a million pounds of rice, in 100-pound bags, made up more than half of the rations. The remainder included a quarter-million loaves of bread, 2,000 barrels of salted codfish, thousands of cans of condensed milk, hundreds of bottles of vinegar, scores of wooden crates filled with tea or coffee, more than a thousand pounds of bacon and a small amount of tobacco. Someone also thought to order 10,000 pairs of pants and 10,000 pairs of shoes to give to survivors. The purchasing contracts for these supplies did have a stringent requirement: All provisions had to be delivered to the Brooklyn Navy Yard and ready for loading onto the *Dixie* within twenty-four hours. As one Army officer who was organizing the procurements said to a newspaper reporter, "We've not been so busy since the last war."

On Tuesday, Secretary Moody issued a press statement that said he had ordered the captain of the *Dixie*, Robert Berry, to reduce his regular crew of about 700 men to less than 200 to make room for the large amount of relief supplies. Also, the statement continued, the *Dixie* would be transporting "a number of newspaper correspondents" and "several scientific men." Only one of the scientific men was mentioned by name, "Thomas Jaggers of Harvard."

The statement went on to say that Jaggar would be carrying "many instruments for recording volcanic movements and apparatus for calculating the peculiar actions and disturbances in Martinique." Jaggar's "many instruments" included a Kodak Brownie camera he had purchased for the trip, an aneroid to determine elevations by barometric pressure, a mercury thermometer that he would use to measure the temperature of steam vents, as well as a tent with camp utensils and an outfit of picks, shovels and hatchets. He also carried a gold watch—a present from his father on his twenty-first birthday—that he would use to time any volcanic activity that might occur while he was in the Caribbean.

The next day, May 14, Moody released the names of the other scientists who would be sailing on the *Dixie*. One was Robert Hill, a geologist for the United States Geological Survey. Hill had tried to enlist in Roosevelt's Rough Riders at the beginning of the Spanish-American War, but was rejected for medical reasons. After the war, he had been sent by the American government to conduct a quick survey of the mining potential of Cuba and Puerto Rico. During the survey, he had found time to travel to several islands of the Lesser Antilles, including Martinique.

George Curtis, who also worked for the United States Geological Survey and had also been sent to the Caribbean after the war, was a cartographer whose expertise was building detailed models of the nation's major cities. His most recent model was of Boston, which, it was said, "was so perfect that it was sent to the Paris Exposition of 1900." Curtis was currently working on one of Washington, D.C., for the United States Senate.

Edmund Hovey was the assistant curator of the geology exhibit at the American Museum of Natural History in New York. He was interested in going to Martinique to collect any curiosities related to the eruption of Mount Pelée that he might find and give them to the museum.

The oldest and most distinguished scientist of the group was Israel Russell. Russell had been one of the founders of the National Geographic Society—established in 1888—and had led the Society's first expedition, a failed attempt to reach the summit of Mount St. Elias, the fourth highest peak in North America, located along the Alaska-Canada border. Russell was now a professor of geography at the University of Michigan and had recently published the book *Volcanoes of North America*, which gave him the credentials to join the *Dixie* on its voyage to Martinique.

Before the team of scientists boarded the *Dixie*, both Hovey and Russell gave interviews to newspaper reporters, saying that, in their opinions, the reports received from St. Pierre about the total destruction of the city and deaths of tens of thousands was almost certainly exaggerated, the same opinion offered by other geologists, most notably by Shaler at Harvard. "When the truth is learned," Russell told the reporters, "you will find that there were 800 to 1000 killed. It is quite clear to me that it was simply a small eruption of one of the volcanoes."

Perhaps because he was the youngest and least experienced of the scientists going to Martinique, Jaggar was more circumspect. Before his departure, while still in Cambridge, he was quoted as saying, "I have no opinions yet regarding the eruptions in the Antilles. I haven't had time to form any. I am simply going to take scientific observations."

He was asked how he had managed to get appointed to the scientific team. He answered quite plainly. "I have moved heaven and earth to go on this expedition and now I have succeeded." The reporter noted that, when Jaggar spoke to him, there was "the embodiment of energy" on his face.

———•———

Because this was his first trip to the tropics, while he was waiting in New York for the *Dixie* to depart, Jaggar bought a pith helmet to protect himself from the tropical sun and a white suit to protect him from the tropical heat. With these in hand, he hurried to the Brooklyn Navy Yard where he would meet the other scientists and board the *Dixie*.

By the time he arrived, the ship was already swarming with newspapermen and other writers. Among them was Carsten Borchgrevink who had led the first expedition to overwinter on the Antarctica mainland. During the expedition, made in 1899, Borchgrevink had seen both Mount Erebus and Mount Terror, the world's most southerly volcanoes. As he prepared to board the *Dixie*, Borchgrevink proudly displayed a pair of asbestos shoes of his own manufacture that he intended to use on the presumably hot ground and scale the heights of Mount Pelée. To his dismay, after arriving at St. Pierre, he would discover that an ordinary pair of rubber boots would have been more useful.

George Kennan was another writer and seasoned traveler who joined the *Dixie*. Kennan had spent years traveling across Siberia documenting the effort to find an overland route for a telegraph line that would link Europe and the United States. He had been in southern Africa during the Boer War. More recently, he had covered the assassination of President William McKinley in the previous September. He was making the trip to Martinique as a writer for *Outlook* magazine, one of the country's major monthly magazines, whose editors asked him to focus on how the eruption had affected the lives of the local people.

The loading of the *Dixie* took longer than expected. Because it was a war vessel, it had no loading boom, and so, every bag of rice, every wooden crate filled with coffee, every bottle of vinegar had to be brought on board individually. At one point, one of the ship's officers heard the clinking of bottles coming from a case marked "Accordions" as it was being carried onto the ship. The officer had the case opened. It was filled with whiskey bottles. He reported the discovery to Captain Berry. Berry called out, asking if anyone wanted to claim it. When no one did, he had the bottles thrown overboard.

The *Dixie* finally pulled away from the dock at half past nine o'clock Wednesday night, May 14, almost a week after the explosion of Mount Pelée. As the ship sailed down the East River and under the Brooklyn Bridge, both civilians and military people gathered on deck to see the lights of the New York skyline. They were dressed in overcoats. There was a quiet stamping of feet to keep warm. Soon, as each man knew, he would be in warm tropical waters.

There were no accommodations on board the *Dixie* for passengers, and so the scientists were assigned quarters vacated by junior Navy officers. The newspapermen and other writers were billeted at the stern of ship over the hold where the barrels of codfish were stored. The constant unpleasant odor soon gave rise to a running joke among those living on "Newspaper Row." It was commonly agreed that, if the wind was blowing in the right direction, the survivors on Martinique would *smell* the *Dixie* long be they could *see* it.

The civilians spent the long days of passage getting acquainted with the ship's daily routine. After breakfast, when the crew turned out for calisthenics, so did the five scientists and most of the journalists, except two portly ones. It was jumps and vigorous knee bends, then three times running around the deck as the ship's band played a quickstep. The five scientists competed among themselves to see who was the fastest. Later in the morning, as the ship's crew ran through fire drills, the civilians lounged in deck chairs under a canvas awning. A giant chessboard painted on the ship's deck with giant iron pieces to move offered some diversion, as did the ship's mascot, a monkey, "General Weyler," acquired during the recent Mediterranean cruise and named for the Spanish general who had been defeated at Havana. For some reason, the monkey was deathly afraid of an empty glove. The crew often gave the monkey a beer that it swigged from a bottle, making the crew wonder how common this was in the jungle.

At night the volcano men challenged the medical men—a dozen Army officers from the hospital corps were on board to treat those who had been injured by the eruption or who might need other medical attention—to a tournament of table tennis, the ship's officers standing around smoking and judging the calls and chasing after an errant ball before it headed out an open doorway and toward the sea. Meanwhile, in the background, one could hear sailors playing guitars and singing the popular songs of the day, such as "On the Road to Mandalay." The most popular tune, which was sung or hummed by almost anyone during the day, was "It Will Be a Hot Time in the Old Town To-night."

On the last night at sea, all of the civilians and most of the officers and crew woke early and stood on deck, shoulder to shoulder, hoping to see some volcanic activity. The morning was warm and the sky was filled with stars. A full moon hung low in the west, its light illuminating a line of distant thunderheads. Finally, the rugged outline of Martinique came into view.

Someone who was familiar with the island directed the attention of the others to the conical profile of Mount Pelée. All was dark. There was a growing concern among those standing on deck that they had arrived too late to see anything interesting when a faint red glow appeared at the top of the volcano, then faded away. A brief cheer arose from the ship.

Captain Berry steered the *Dixie* closer to the island. The men could now see two faint red glows at the base of the volcano where the island met the sea. Someone called out, "Is that the eruption?" Someone else answered. "No, that's where they're burning the bodies."

With that brief exchange, the temper of the voyage changed. The days of lightheartedness had ended. Those onboard the *Dixie* now realized that they were about to face life's grimmest reality—the possibility of death.

———•———

At seven o'clock, the sun now high above the horizon, the *Dixie* steamed slowly into the harbor at Fort-de-France, twenty miles south of St. Pierre. Several warships were at anchor. Each one had maintained full steam since its arrival days earlier in case it had to put to sea at a moment's notice.

The United States Navy cruiser *Cincinnati* had arrived twelve days ago, two days after the eruption. Other cruisers in the harbor were the French Navy *Suchet*, the British *Pallas* and the Dutch *Koningen Regentes*. All four ships had unloaded their extra stores and had sent medical teams on shore. Also at Fort-de-France was the sea-going tug *Potomac* of the United States Navy. It had been at San Juan, Puerto Rico, at the time of the eruption and had received orders to sail at once to Martinique. Since its arrival, the *Potomac* had been making trips around the islands rescuing survivors.

As soon as the *Dixie* anchored, an officer from the *Cincinnati* came alongside in a launch and shouted to Captain Berry that new orders had arrived from Washington, D.C. Enough supplies, so the orders stated, had been delivered to Martinique. Berry was to sail the *Dixie* south to the island of St. Vincent and give out supplies and provide medical aid to victims of the eruption of Soufrière.

Berry gathered his passengers and told them that he could give them only one day at St. Pierre. After that, each man would have to decide for himself whether to remain on Martinique, and find his own way back to the United States, or continue on the *Dixie* and sail to St. Vincent.

At 10:30, the *Potomac* pulled up next to the *Dixie* and took onboard those who wanted to go to St. Pierre. A party of officers from the *Cincinnati* and from the *Dixie*, as well as most of the journalists and all five American scientists, transferred to the *Potomac*, almost overloading it.

The tug cruised north, keeping close to shore. From what those on the *Potomac* could see, the upper reaches of the volcano were still covered with dense vegetation. Lower down were fields of sugar cane ready to be cut and along the coast fringes of coconut palms swaying in the sea breeze. Nowhere was there any sign of a disturbance from the volcano. Occasionally, a village was spotted comprised of a few whitewashed houses and a parish church with a high spire. People were usually nearby, standing and waving at the boat.

As the *Potomac* proceeded, headland after headland was passed. Each one was the truncated end of a sharp-crested ridge that ran from near the volcano's summit to the sea. After an hour of travel, the *Potomac* rounded yet another headland, seemingly like the others. Though, this time, as Jaggar remembered the scene, in front of him "burst into view one of the most weird, spectral sights I ever expect to witness."

The changed occurred immediately. It had the precision of a knife edge. To the south was dense green foliage. To the north and continuing as far as Jaggar could see was a featureless plain of drab browns and grays. Later inspection would show that in some places the change was so abrupt that it was no wider than the trunk of a single tree, one side green and the other scorched brown.

The plain itself was completely barren—no indication that, just two weeks before, this had been one of the most popular ports in the Caribbean. There was no evidence of the recent hustling of people. The only movement came from short lines of steam that crisscrossed the plain and where, along the shoreline, small balls of steam were rising passively into the air.

"St. Pierre is over there," said Lieutenant Benjamin McCormick, in command of the *Potomac*, pointing to where muddy surf was breaking against a bleak and sandy shore. Jaggar looked in vain for a pattern of streets, but all he could see were vague outlines of a few low walls that marked where St. Pierre once stood.

A few rusted metal buoys were still bobbing close to shore. The crew tied the *Potomac* to one. Then small boats were made ready to land those who wanted to go ashore.

Lieutenant McCormick had walked through St. Pierre the previous day, leading members of his crew on a search of the ruins, hoping to recover the bodies of the American consul and his wife, Thomas and Clara Prentiss. But Mount Pelée had exploded again, sending McCormick and his landing crew racing back to the shoreline. Some people later said it had been larger than the explosion on May 8. Fortunately, no one was injured, but, now, the next day, McCormick was obviously nervous.

"There is danger today," he told those who were entering the small boats. "The mountain is behaving badly." He instructed the members of each landing party that if the ship's whistle should blow, they were to run for their boats and return to the *Potomac* immediately.

Jaggar and his party were put ashore at a stone jetty near the center of the now-dead city. With no streets to guide them, they wandered about, stumbling and falling over rubble and ash heaps. Many walked away in groups of twos or threes. Jaggar set off alone.

Not one single building had survived. The heavy iron sheets that had once served as roofs had been blown away and were found crinkled and bent like cardboard. Only a few walls, none higher than a person's chest, were still standing. And everything was covered by a thick layer of grayish volcanic ash wetted by rain. At one point soon after he stepped on shore, Jaggar stopped and looked at his feet. His boots were covered with slippery ash, which made them much heavier than usual. He bent down and ran the ash between his fingers. He smelled it. It had the tinge of sulfur. It reminded him of damp gunpowder.

That there had been a furious wind during the eruption was indicated by bent and twisted rods of iron and by huge trunks of trees that had been torn from the ground. Old cannon used as mooring posts at the quay had been torn up and moved. At one place close to the shoreline, a three-ton iron statue of the Virgin Mary had been swept off its pedestal and moved more than 50 feet. At a distillery, huge tanks made of quarter-inch boilerplate were peppered with holes where rocks had struck and penetrated, giving the impression the tanks had been bombarded by artillery.

And dead bodies were everywhere. Jaggar found a dead baby in an iron cradle and an old man face down in an empty water tank. At a bakery, he opened the metal doors of a huge oven. Inside were the remains of a man who was found laying on his back, his head on one arm and his legs pulled up under him. The man's body was so thoroughly cooked and shriveled that his thighs were only a few inches thick. The heat had drawn his skin away from his knees and elbows, exposing the bone. Apparently, the man had climbed into the oven to protect himself when the volcano exploded. Ironically, he had roasted inside his own oven when it was surrounded by hot volcanic ash.

Jaggar continued to wander through and search the ruins. He occasionally joined others. Deeply disturbed by what he was seeing, he would describe the scene as: "The air was heavy with a haunting odor that one dreams about afterwards; it is a combination of foundry and steam and sulphur matches and burnt things, every now and then a whiff of roast, decayed flesh that is horrible."

All life had been extinguished. After years of academic study, Jaggar had never read an account of such sudden and complete fiery destruction—not in war and not by nature. Then he remembered. He had seen this before.

———•———

St. Pierre was a modern-day Pompeii. The two cities had remarkable parallels. Both had been thriving seaports with a lively commerce of exports and both were filled with hotels and restaurants that catered to tourists. The cobblestones on the streets and the layout of buildings with open courtyards and walls of painted plaster were also similar. And the fate of each city had been decided by the sudden explosion of a volcano that had been quiet for many years.

Two thousand years separated the two disasters—which made Jaggar wonder: Why hadn't human knowledge advanced enough since Pompeii to prevent a replay of such suffering?

It was the beginning of the 20th century, and, to Jaggar and many others, science and engineering seemed to be the keys to progress. Already a vast network of railroads crisscrossed the nation, the largest such system in the world. And railroads were becoming faster and more efficient all the time. So were ships, which were larger and faster, and more reliable and safer than ever, designed to ease travel and increase commerce around the world. There was even popular support in the United States for the building of a canal across Panama that would connect the Atlantic and Pacific oceans. In short, it was an age when all things seemed possible.

But what of the horror at St. Pierre? Some scientists were already talking about predicting and controlling the weather. In fact, a weather bureau had been established in the United States the previous decade and was issuing daily forecasts. Giant dams were being constructed to control the flow of the nation's largest rivers, thereby ending the threat of floods. Likewise, so Jaggar thought, the fear and destruction brought by volcanic eruptions could—and should—also be alleviated. It should be possible to predict and control such natural phenomenon. What was needed was a better understanding of how volcanoes work.

———•———

Late that morning, Jaggar was on a hilltop outside the city studying the volcano through binoculars. Clouds shrouded the summit. He examined the lower slopes. He noticed wisps of steam rising from several places. As he continued to watch, the steaming became more vigorous. Then more places became active. Eight, ten, then twenty places were emitting steam. Now there were forty scattered across one side of Mount Pelée.

He looked toward the *Potomac*. Captain Berry had apparently seen the same thing. The ship's whistle blew. Jaggar looked back at the volcano. More steam was appearing. And it was becoming more vigorous. In fact, in some places, it seemed to be jetting out of the volcano. Jaggar saw people running for the shoreline. One landing boat had just shoved off when two more people appeared. It went back to pick them up.

Now Jaggar began to run. Then he stopped. He realized that what he was seeing was the coolness of a sudden rain shower falling on pockets of hot volcanic ash, causing steam to rise, but, at the time, to a group of inexperienced volcano watchers, including himself, it seemed as if the entire volcano was breaking apart and was about to explode.

———•———

The next day, the *Dixie* left Martinique and sailed for St. Vincent. Jaggar and two other American scientists, Hovey, the museum curator, and Curtis, the cartographer, were onboard. The other scientists had decided to remain on Martinique.

At St. Vincent, Jaggar and the others met Mr. T. MacGregor MacDonald, a local sugar planter, who had witnessed the May 7 eruption of Soufrière and agreed to guide the three Americans to the volcano's summit.

MacDonald hired six islanders to take them by dugout canoe from Kingston, the main town on St. Vincent, to the mouth of the Wailliabou River, a distance of about twelve miles. They made the trip at night, arriving at the mouth of the river at daybreak. The volcano's summit was shrouded in a light mist. They started an ascent immediately.

The six islanders worked as porters, carrying instruments, water and food. MacDonald led the way, following what remained of an old trail, first passing through dense jungle, the vegetation heavy with volcanic ash, then across a landscape desolated by the May 7 explosion.

Progress was slow, made harder by a strong headwind that blasted the men with volcanic sand, forcing them to squint, making it almost impossible to see. Near the summit, the men slogged through knee-deep wallows of mud. They climbed over fields of huge jagged rocks recently blown out of the volcano. At last, after three hours of tortuous climbing, they reached a crest and were within sight of the summit.

A thick fog surrounded them, blocking direct sunlight. The six islanders decided to retreat. The three Americans and MacDonald remained. After another hour, the fog cleared and the four men crept upward toward the edge of the summit crater.

At their feet lay a vast opening. The inner walls were sheer precipices of freshly broken rock. On the side opposite them, in clear view, rose a huge chimney of roaring steam. And below, where the steam originated, was a lake of pale green muddy water.

The surface was bubbling at a hundred points. A small trail of steam rose from each point, the trails coalescing halfway up the wall into the chimney of steam.

The men set to work scribbling in notebooks, taking photographs and getting dimensions of the crater. Jaggar began by marching off a baseline. Then, with a compass, he measured several angles to distinct points, determining that the crater was more than a mile wide and that the depth down to the lake was more than half a mile. He took out an aneroid and, comparing his reading to the elevation on an old map, decided that at the spot where he was standing, the eruption had blown away the upper thousand feet of the volcano.

Before they left, the four men built a stone cairn to indicate to anyone who may come later that they had been first.

Then the four men retraced their steps and went down the volcano. As they passed through the village, Jaggar noticed that women were bringing out their children, so he thought, "to gaze at us, the godlike men who had dared the crater." It was his greatest performance yet.

———•———

After five days at St. Vincent, the *Dixie* sailed away. Jaggar and his two American colleagues stayed behind. They continued to explore the island and attempted another ascent of Soufrière, but poor weather forced them to retreat. Their conclusion, after comparing notes with those who had stayed on Martinique, was that the explosive eruption of Soufrière on May 7 had been more energetic—that is, had blasted away more of the volcano's summit and spewed out more volcanic ash—than the May 8 eruption of Mount Pelée that had leveled St. Pierre. The greater destruction on Martinique had been because of the unfortunate direction of the blast that had swept over St. Pierre and had caused so many horrible deaths.

After three weeks on St. Vincent, Hovey and Curtis found passage back to the United States. Jaggar decided to stay longer in the Caribbean. He began his extended stay by sailing to several islands to collect samples of volcanic ash that had fallen and to measure the thickness. One of the islands was Barbados. While there, he learned that two survivors of the *Roraima*, which had been at St. Pierre on May 8, were convalescing at a nearby hospital.

A dozen passengers had been on the *Roraima* when Mount Pelée exploded. These included four members of the Stokes family and their private nurse, Clara King.

The mother, Clement Stokes, had been living in New York, but was recently widowed and had decided to return to Barbados and live with her sister where she could take better care of her children, who were traveling with her. The oldest was nine-year-old Marguerite. A son, Eric, was four. The youngest was Olga who was three.

On the fateful morning, shortly before eight o'clock, the nurse, King, left the others in the stateroom and stepped outside onto the deck. She saw others gathered along the railing and pointing at the volcano. She heard the ship's steward call out that something was happening at Mount Pelée. King went to the railing and looked. She saw a black cloud surrounding the summit. Then, seconds later, the cloud exploded and swept down the volcano toward her. Within minutes, it had passed over St. Pierre and had hit the sea.

King managed to reach the stateroom and close the door just as the cloud hit the ship broadside. She felt the *Roraima* being lifted high, then rolled over and dropped down.

King and the others were thrown off their feet. As the ship righted itself, it bobbed as if in a rough sea. Next came darkness, then intense heat and the feeling of suffocation. A skylight broke and hot ash poured into the stateroom and down on the occupants.

Sometime later, King would never be able to remember how long, the door of the stateroom burst open and a feeble ray of sunlight illuminated the room. King could see Clement Stokes and her son huddled in a corner covered with scalding mud. The infant girl was dead. Marguerite, who had managed to find King in the darkness, was sitting next to her, clinging tightly to the nurse. At one point, the young girl put her hand down to move closer. As she did, her arm plunged up to the elbow into hot ash that burned away much of the skin.

They left the stateroom and made their way on deck, Clement Stokes carrying the dead infant. Other survivors had already gathered. The lifeboats had been swept away and so several men were trying to build a raft of loose timbers. Someone found a blanket and placed it around the Stokes boy. His hair and clothes had been burned off. He died soon after.

The ship was on fire and the city was in flames. Clement Stokes turned to the nurse and gave her what money she had and asked King to promise to take care of the young girl. Then the mother died.

The fall of red-glowing ash and cinders began to lessen slightly. King saw a ship approaching. It would later prove to be the *Roddam*, captained by Edward Freeman. King thought the ship was coming to rescue them, but the *Roddam* swerved and headed out to sea. As it passed close to the *Roraima*, several survivors from that ship jumped into the water and swam for the *Roddam*. They were pulled under the water by a dark and turbulent sea.

For six hours, King and the others clung to the wreckage of the *Roraima*. Finally, a ship arrived and rescued them.

Jaggar met King in her hospital room. She told him of her ordeal. Jaggar could see that she had bandages around both knees and along one arm. Also in the room was Marguerite Stokes. Burns had left deep scars on her head and along both arms and on her hands. One ear was disfigured. Jaggar knew she would be crippled for life.

The night before meeting King, Jaggar prepared questions to ask her. How had the plume rising above the mountain appeared just before the explosion? It was gray, mostly white below and black above, she answered, with smoke rolling to the left. She said the sea had been calm before the eruption and that she heard no explosions, just a distant rattling of thunder.

Of all the things that Jaggar had seen and would do in the Caribbean that summer, it was the meeting with Clara King that made the deepest impression on him. Here was a small frail woman destined for a life of servitude. And she had seen—and survived, at close range—a major volcanic explosion. It was probably more activity than he would ever see.

"It was the human contacts, not field adventures which inspired me," he wrote years later when he recalled the meeting. "Gradually I realized that the killing of thousands of persons by subterranean machinery totally unknown to geologists and then unexplainable was worthy of a life's work."

And so he was decided. He would devote himself to a study of volcanoes. It was an ideal—and ideals come at a price. It would take a few more eruptions—and another decade—before he was completely willing to pay his price.

CHAPTER FOUR

CHAMPAGNE

The unexpected appearance of a new volcano in 1831 almost threw the global political world into turmoil.

On July 11 of that year, in the central Mediterranean, Captain Charles Henry Swinburne, in command of the British sloop *HMS Rapid*, saw an unusual column of white clouds rising from the sea. He went to investigate.

Four hours later, now night, Swinburne was close enough to see that, at the base of the column, hot, cherry-red embers were being flung into the air, then falling back into the sea. The column itself was filled with flashes of lightning. Swinburne held the *Rapid* on station for eight days and nights, he and his crew witnessing a rare sight—the birth of a new island.

Because no land had been here before, instead of sailing on, Swinburne returned to the headquarters of the British Mediterranean

Fleet at nearby Malta to report the discovery to the vice-admiral. The vice-admiral sent out another ship, the *St. Vincent*, with men and equipment to survey the new island and determine its exact position. In less than a day, it was determined that the island was a third of a mile across, nearly round in outline, and consisted of two prominent hills at opposite ends of the island that were connected by a low area that was barely above sea level. At the top of one hill the Union Jack was raised. With this small act, the vice-admiral claimed the new land for England, naming it Graham Island after Sir James Graham, the First Lord of the Admiralty.

The Sicilians arrived next, on an appropriately named ship, the *Etna*. Its crew removed the British flag and replaced it with one of their own. They also rechristened the island Ferdinandea, in honor of the Sicilian king, Ferdinand II. Then the *Etna* sailed away.

A French ship arrived. On board was Constant Prévost, a member of the Académie des Sciences in Paris. He spent several days on the islands. On one of those days, he watched as officers of the French Navy lowered the Sicilian flag and replaced it with the Tricolors, Prévost commenting that the effort was "a vain and ridiculous ceremony." The French also rechristened the island, naming it Guilia because it had been discovered in the month of July. Then the French sailed away. And thus the political wrangling began.

The sea passage between Sicily and North Africa is of strategic importance because it links the broad eastern and western sections of the Mediterranean Sea. During the 19th century, the Sicilians controlled the area north of the passage, the French controlled the area to the south off the coast of Africa and the British were in the middle on the island of Malta. The appearance of a new island meant someone might use it to build a new navy base, upsetting the balance of power. And so each government began to make contingency plans. In the end, nature had the final word.

By winter—the eruption had stopped in August—storms battered the small island, and it was rapidly pulled back under the sea. By the end of 1831 a rocky crag was all that was left. By the end of January 1832, even that was gone.

As a footnote, when the British landed and raised the Union Jack, someone thought to fill a few buckets with the new volcanic material. That material is stored today in the Natural History Museum in London, all that is left, so some cynics say, of the smallest island ever ruled by Britannia.

But the island did leave an important scientific legacy. When Prévost visited, there was a small lake of reddish water contained within a circular basin at the center of the island. He took some care in examining the lake, noting that bubbles of gas were rising from the bottom and bursting at the surface. Each tiny burst sent out a small spray of water. It reminded him of what happens when a bottle of French champagne is opened. From that analogy, Prévost suggested that volcanic eruptions were powered by the escape of gas bubbles dissolved in molten rock. On reflection, it was a momentous suggestion—but one that would be forgotten for more than eighty years, eventually revived by those who were present to witness the continued eruptions of Mount Pelée.

———•———

While still in Barbados, Jaggar wrote a letter to *The Boston Globe*, describing his Caribbean adventure up to that time. He included in the letter the standard explanation for explosive eruptions, that is, that they were the product of massive steam explosions. But he must have doubted whether this explanation was adequate to explain what he had seen at St. Pierre because, at the end of the letter, he informed the people of Boston that he was delaying his return because he wanted to return to Martinique and determine, as best he could, what had powered the explosion that had leveled St. Pierre and had caused so many deaths.

All of his American colleagues had departed Martinique when Jaggar returned on June 24. Two weeks later, on July 9, he was in Fort-de-France and, as he would recall, "I had the good fortune to see an eruption of the first magnitude on Mount Pelee."

It was 8:20 in the evening and the sky was darkened by twilight. He was inside a local library when someone entered and, recognizing him, told Jaggar that something extraordinary was happening on the volcano. He rushed to look.

People were already in a panic on the street. Looking in the direction of Mount Pelée, Jaggar could see a column of black clouds billowing up to a great height. Within minutes, the top of the column had spread out and covered most of the sky.

Jaggar hurried to a nearby hotel and climbed to the roof to get a better look. The night air was still. He heard only the hum of tree frogs and crickets and a faint low roar coming from the volcano. He stood and waited. Soon a fantastic light show—one that he would later say he had never thought was possible—began.

On the edge of the eruptive cloud were shimmering points of white lights. As time progressed, these points grew into bolts of lightning that passed into and out of the cloud, the flashes becoming more frequent and grander as the eruption progressed. At one point, about an hour after the display had begun, the whole cloud was filled with dancing lights. And then there was the thunder.

At first, the rumbling overhead could barely be heard. But that soon evolved into sequences of thunder rolling across the sky. A sequence began as a bubbling growl that grew into a crescendo. The pattern of lightning also changed. At the height of the display, lighting shot out of the eruptive cloud as long serpents, then as sheets, illuminating the entire sky. One remarkable flash made a complete loop, arching out and down from high on the cloud, touching the top of the volcano, then curling back up again.

The display of lightning and thunder lasted two hours. At daybreak, anxious to see what had happened nearer the volcano, Jaggar hired a small boat to take him to St. Pierre.

When he arrived, he discovered that the explosive eruption of the previous night had covered the ruined city with yet another blanket of volcanic ash. And there was something else that he encountered that day at St. Pierre. He met two other scientists who were also inspecting the ruins. They had recently arrived and had seen the previous night's eruption at extremely close range.

Responding to the May 7 eruption of Soufrière on St. Vincent—the island was then a British colony—the Royal Society of London had sent two scientists to investigate. One was a noted professor of geology at the University of Edinburgh in Scotland, John Flett. The other was a self-described "amateur of limited leisure," Tempest Anderson. Anderson had been a physician who specialized in diseases of the eye. His interest in the human eye led him to study the new technology known as photography. (He made his own cameras and lenses.) And photography led him to search out and document what he regarded as nature's most magnificent force—volcanic eruptions.

His introduction to volcanoes came during a visit to the Eifel region of Germany where there are many extinct volcanic craters filled with picturesque lakes of water. That led him to see volcanoes in action, first to Vesuvius in 1883, then Etna the same year. Both volcanoes were in states of minor eruptions, small cones on each one spewing out small rains of cinders. From that time on, Anderson became a volcano-chaser, probably the first one in the world, dedicating himself to traveling at a moment's notice to see and record, by writing descriptions and taking photographs, the great volcanic eruptions of the age. In fact, at his home in Edinburgh, he kept two suitcases packed, one with clothes for a cold climate and the other for a hot one, just in case he received news of an eruption somewhere in the world and he had to leave in a hurry.

After recording the destruction by Soufrière on St. Vincent, Flett and Anderson sailed to Martinique. They were relaxing on a small sailboat, bobbing in a windless sea, one mile off the coast of St. Pierre on July 9 when Mount Pelée exploded.

The two men happened to be discussing the possibility of climbing the volcano the next day when they saw a yellowish glow reflected off clouds hanging over summit. "It was like the lights of a great city on the horizon, or the glare of large iron furnaces," Flett and Anderson wrote later.

Then, from the volcano, they heard a "prolonged angry growl." Instantly, the small cloud tumbled down the side of the volcano leaving a bright orange streak. The speed, as both Flett and Anderson would later attest, "was tremendous."

"It was a fear-inspiring sight," wrote Anderson, "coming directly for us, where we lay with sails flapping idly as the boat gently rolled on the waves of the sea."

The boat's crew, however, had been in a panic since seeing the initial yellowish glow. The men seized oars and rowed for their lives, but the boat was too heavy and awkward to move that way. Several members of the crew fell to their knees and started to pray. Flett and Anderson continued to watch, "overwhelmed by the magnificence of the spectacle."

Just as the rapidly descending cloud seemed ready to engulf them, the cloud's front lost its downward momentum and rose into the air, passing directly over the boat. Now a hail of pebbles, some as large as chestnuts, began to fall on deck and into the sea. And that was followed by a rain of fine ash that lasted for more than an hour. Flett and Anderson realized they had witnessed at close range the deadly monster that had leveled St. Pierre.

———•———

I can attest to the accuracy of their account because I saw a similar glowing cloud at a much later time at a different volcano. It happened in the spring of 1982 at Galunggung volcano on the island of Java.

My Indonesian colleagues and I were three miles from the volcano. It was evening twilight and the eruption began, as it had for Flett and Anderson, with a faint yellowish glow reflected off a cloud hanging low over the volcano. Within seconds, a faint roar was heard. Next the glow started to pulsate, rising and falling in intensity. After a few cycles, a single explosion blew red and orange streamers into the air, their trajectories resembling the arcs made by ribs of an open umbrella. Darkness followed.

Minutes passed. The glow returned to the crater, this time brighter than before. The roaring was louder. Then, as the glow continued to brighten, a flood of incandescent material, bright orange in color, sloshed up and over the crater rim. We watched in awe as it poured down the side of the volcano and toward us. We lost sight of the front as it passed on the opposite side of a low intervening ridge. The whole length of the ridge was silhouetted in light. It dimmed slowly.

We were in darkness a second time. The roaring of the volcano continued, punctuated by deep detonations. Then there was another burst of the volcano, sending red and orange streamers higher than before. And from that rose a fountain of incandescent rock to heights of thousands of feet. And above the fountain was a pillar of smoke, the edges outlined by a clear night sky.

The lightning and thunder display came next. It started directly over the volcano, and then spread across the sky. It was like standing under a bridge while huge trucks rolled overhead, sparks from the wheels sending out bolts and sheets of lightning.

Next a hail of stones began to fall, followed, minutes later, by volcanic ash. It blocked out our view of the volcano. All we could hear was the roar.

I had decades of work by other investigators to rely on when I saw my spectacle. Jaggar and Flett and Anderson were trying to understand a deadly phenomenon that, in their era, had barely been hinted at.

What had propelled the glowing cloud down the side of Mount Pelée? Jaggar and Flett and Anderson attributed it totally to gravity, Flett and Anderson having described the movement of the cloud as similar to "a toboggan on a snow slide." They were partially right. Gravity does play a major role in guiding such a cloud down the side of a volcano. But what of the tremendous speed? The cloud that had destroyed St. Pierre on May 8 had traveled four miles in about two minutes, an astonishing speed of more than a hundred miles an hour.

The answer to this more difficult question came soon. It came from someone else who arrived at Martinique to see the ruins of

St. Pierre. But he would not uncover the answer until he made a grievous mistake.

———•———

Alfred Lacroix had risen quickly to a position of prominence in French science. Having graduated with a doctorate degree in mineralogy in 1889 from the Muséum d'Histoire Naturelle in Paris, just four years later he was named the head of the mineralogy section at the museum. His meteoric rise was due, in part, to an extraordinary ability to study and describe and reveal the physical and chemical properties of minerals. And to the fact that he married the museum director's daughter.

The father-in-law, Ferdinand André Fouqué, is still regarded as one of France's most accomplished geologists. In 1866 he had gone to Santorini in Greece after that volcanic island exploded. While there investigating the recent activity, he discovered an ancient city buried beneath more than a hundred feet of volcanic stones, cinders and ash erupted by previous eruptions. Those ruins are now thought to be the ancient city of Atlantis, a place that, according to the Greek philosopher Plato, "after a single day and night of misfortune" sank into the sea.

As soon as news of the 1902 explosive eruption of Mount Pelée reached France, Fouqué asked his son-in-law to prepare a lecture about the geology of Martinique for members of the Muséum d'Histoire Naturelle. The lecture was presented on May 26. A week later, the first samples of new ash erupted from Mount Pelée arrived in Paris. Fouqué gave the samples to Lacroix who studied them for a week, who then gave a second lecture to museum members about what he had discovered. A third lecture was scheduled for late June in which Lacroix would compare his study of the 1902 samples with samples Fouqué had collected on his travels years earlier of the 1851 eruption. But, before the third lecture could be given, Fouqué raised a concern that, up to that time, only a team of American scientists was studying what had happened at St.

Pierre. Martinique was a French colony. French science should be presented. And so the third lecture was postponed indefinitely. And Lacroix was sent to Martinique.

He and two assistants arrived on Martinique on June 23. They tramped through St. Pierre, taking photographs. They interviewed survivors who, by this time, had told their stories dozens of times. In short, the French repeated what had already been done. But Lacroix did have the foresight to train two local French soldiers on the use of cameras and to station them on a small hill south of St. Pierre where there was a good view of the volcano. He drilled them endlessly on how to take a quick sequence of photographs until the two soldiers could take a photograph every ten seconds. Lacroix then left to tour other Caribbean islands and study their geology.

On July 9, the two soldiers failed to respond fast enough to capture the beginning of the eruption on film, but they did succeed in photographing the ash cloud that sped down the volcano and toward St. Pierre. When Lacroix returned, he examined the photographs, seeing, for the first time, documented proof of the high speed of the cloud.

On August 1, Lacroix departed the island and sailed back to France. Before he left he met with the French colonial governor and gave him his personal assurance that Mount Pelée was lessening in explosive activity and that the worse was over. That caused the governor to issue a directive that ordered everyone who had left their houses and were living in Fort-de-France—the number of refugees was in the tens of thousands—to return home or risk losing their food allowance. And so, many returned. Jaggar, learning of the directive, issued his own appraisal of the risk.

"The mountain at this moment appears calm," Jaggar wrote in a report that he sent to the colonial governor, "and the dust columns that one sees from time to time are largely due to landslides from the crater." But, he cautioned, future eruptions, including a repeat of July 9, should be expected. Furthermore, he said that he was "strongly opposed" to the island's government forcing people to return to their homes when, in his opinion, "the volcano is still dangerous."

Jaggar also identified the area of highest risk. "I do not think a single habitation northwest of the line from Bellafontaine to Basse-Pointe area safe." That area covered almost the entire western half of Mount Pelée, which included St. Pierre. In addition, he thought towns located high on the volcano, such as Morne-Rouge, "are [not] safe at present." At the end of the report, Jaggar rued the colonial government to establish "a properly equipped experiment station," one that would have a permanent staff of scientists who could warn people of changing activity and imminent eruptions. Then, a few days after he released his report, Jaggar left and returned to Boston.

Three weeks later, Mount Pelée exploded again.

———·•·———

On August 25 an enormous cloud of ash spewed out of the volcano, the ash falling across the island. Many of those who had recently returned to their homes now retreated back to Fort-de-France. Some in Morne-Rouge—one of the places Jaggar had identified—stayed, seeking protection inside a stone church, the largest and sturdiest building in the small town, where Father Francis Mary welcomed them.

The next day, Father Mary sent a delegation of citizens to Fort-de-France to ask the colonial governor for permission for the people of Morne-Rouge to leave and go to a safer place. The governor said no.

On August 29, the ground at Morne-Rouge shook continuously. The next day, Mount Pelée exploded yet again, devastating an area twice as large as the one destroyed on May 8. During the explosion, Father Mary stood outside his church, holding the door open so that people could enter. An ash cloud, similar to the one that destroyed St. Pierre, swept over Morne-Rouge. Thousands died, including Father Mary.

There were no recriminations within the colonial government in Fort-de-France or at the colonial office in Paris over the decision to deny permission for the people of Morne-Rouge to leave. Lacroix, however, may have felt a personal responsibility because, soon after he heard of the additional deaths, he sailed back to Martinique. This time he would stay six months—time well used in that he would finally determine what gave volcanic eruptions so much explosive force.

The first person to see the odd feature on Mount Pelée was Thomas Jaggar. It was July 6, and he was back in St. Pierre, again surveying the ruins. For days, the summit of the volcano had been hidden by clouds, but on this morning, as he recorded it, "I had the good fortune to see the whole cone clear of clouds for about ten minutes." During that brief time, he studied Mount Pelée through binoculars, seeing "a most extraordinary monolith" at the summit.

It was shaped like the dorsal fin of a shark, smooth and slightly curved on one side, steep and with a slight overhang on the other. Three days later, after the July 9 eruption, he studied the summit again through binoculars. The moonlight was gone. But it reappeared a few days later.

The second monolith was also destroyed, on August 30, by the same explosive eruption that sent an ash cloud over Morne-Rouge. After that, the intensity of explosions lessened, allowing a third monolith to grow. It was soon known as "the spine of Pelée."

It reached its broadest and highest dimensions in the spring of 1903 when the base was 500 feet across and the top towered nearly a thousand feet above what had been the summit. Seen during the day, it appeared as a great steaming obelisk. At night, it glowed a dull red. One captain whose ship passed it at night remembered it as "resplendently luminous." But to the people of Martinique it was an eerie reminder of the tens of thousands of people who were dead. And for Lacroix it was the key to understanding why volcanoes explode.

Lacroix returned to Martinique on October 1, focusing his attention on the growing monolith, studying it at close range. In his field notes, he recorded that the great obelisk changed color depending on the viewing angle. From one side it might appear deep brown, while from others pink or purple or nearly white. He measured and photographed it. At one point, comparing measurements on two successive days, he determined the top of the monolith was rising at the astonishing rate of forty feet a day.

As the spine of Pelée grew, it maintained its original basic shape, as described by Jaggar. One side was smooth, almost polished in places. It had long grooves, some of which began at the base and continued upward for hundreds of feet. It was the grooves that showed how malleable the surface must be. Lacroix likened them to the grooves one gets when applying thick paint.

The other side was rough and had overhangs produced by frequent rockfalls. Occasionally, a rockfall was huge—and Lacroix witnessed several; he was frequently at the summit. And when such events happened, the entire side of the spine would collapse. Then a jet of hot gas would come roaring out from the base of the spine and send a mixture of ash and stone hurling part way down the volcano. It was this sequence that gave Lacroix the insight into volcanic explosions.

As he described them, the ash clouds were "an intimate blend, a sort of emulsion, of solid materials suspended in water vapor and gas, carried together in a high temperature." At the front of a cloud, as he observed in person, was a protruding tongue that always kept close to the ground. As he watched the tongue move, he realized it was not hurling down the volcano simply under the pull of gravity, but was being propelled forward by the constant expansion of gas—gas that had been dissolved in the rock at the base of the spine, then released suddenly during a collapse.

In short, as Lacroix suddenly realized, the spine worked as a cork keeping a shallow batch of hot rock under pressure. When a side of the spine collapsed—that is, when a section of the cork was removed—pressure was released and gas bubbles formed instantly. Lacroix said it was much like the opening of a bottle of champagne—the same analogy Constant Prévost had used, though there is no evidence that Lacroix knew of his countryman's earlier work.

Suddenly, much became clear. Volcanic explosions and the menacing ash clouds they could produce—a phenomenon that Lacroix described using the French term *nuée ardente*, literally, "glowing cloud"—were powered by the expansion of gas bubbles from gas-charged molten rock within the volcano—and the earth—itself.

But to think that Lacroix's insight was the last word in understanding how volcanoes work is like imagining that Benjamin Franklin's flying of a kite during a thunderstorm revealed the fundamentals of electricity.

The lesson of St. Pierre was just a beginning.

CHAPTER FIVE

VESUVIUS

While Jaggar was in the Caribbean, his friend and colleague, Charles Palache, wrote to him, keeping him aware of developments at Harvard.

In the first letter, dated three days after the *Dixie* had sailed from New York and was still en route to the Caribbean, Palache mentioned that Professor Shaler had come into his office soon after Jaggar left to say that he no longer had anything against Jaggar going to Martinique, though he still considered the trip unnecessary. In fact, Shaler had sat in Palache's office for over an hour, reminiscing about a trip he had taken many years earlier to Italy. During the trip he had climbed Vesuvius—he was there in the winter of 1882 when, as he remembered it, "it was possible to creep up to the very edge of the crater and look down upon the surface of boiling lava"—and, in that one view, as he told Palache, he had probably seen more volcanic activity than Jaggar would at Martinique.

In another letter, dated three weeks later, Palache reported that he had finished teaching Jaggar's classes, which included giving the examinations. Palache also reported that the fieldwork they had planned for that summer in the Bradshaw Mountains had been cancelled. Then, perhaps, as an afterthought, he attached a note saying that his cousin from San Francisco, Helen Kline, was visiting and that she would be staying an unknown length of time.

Helen Kline was the twenty-two-year-old daughter of George Washington Kline, a vice-president at Crocker-Woolworth National Bank, one of San Francisco's major banks, a position that allowed his family to live in great comfort.

The family home was a large house located at the intersection of Fillmore Street and Pacific Avenue in one of San Francisco's most fashionable districts. The house itself, as Helen would remember it, was "of grey gingerbread elegance, three stories high with a basement and an attic and a prominent round turret." It set at the center of three adjacent lots. Wide lawns surrounded it on all sides. Passersby could see it through a spiked iron-wrought fence. It was within these confines that Helen spent her childhood, one of wealth and privilege, her every need met by a small army of maids and other house servants. Through it all, as she later lamented, "I grew up bored."

"I can't remember anything special happening to me," she would write in a memoir that she secreted away and was found years later by her daughter, "except that I didn't die."

She also grew up defiant.

> I was always very naughty—a terrific liar, stubborn as the dickens and always saying and doing unexpected things which had to be suppressed . . . I was spanked and spanked, had my mouth washed with various household commodities such as brown soap and mustard—and so far as I can see nothing ever availed.

One of the rare pleasures of her childhood came after the birth of her sister Eliza. Helen was sitting in a large chair at the age of five, grinning widely, when the baby was placed in her lap. It was a moment that gave her "an awful lot of joy." But, to her regret, as time progressed, she and Eliza grew apart. Her sister became an energetic and likable person who was curious about everything and who showed a genuine interest in people. Helen, in contrast, was given to fits of anger and was hard to tolerate. She once tormented a piano teacher to such a degree that the teacher called out, "Mrs. Kline, will you come here and see if *you* can do something with this child." The teacher soon left and never returned.

In the spring of 1902, Helen Kline decided to demonstrate her independence by taking a trip. She announced to her family that she was traveling by herself to Boston. No one challenged the decision.

In Boston, Miss Kline presented letters of introduction to the social elite who, in turn, invited her to afternoon teas on Beacon Hill and to other exclusive affairs. She also called on her cousin, Charles Palache, who invited her to family dinners. It was at one such dinner that she met Thomas Jaggar.

Jaggar was a frequent guest at the Palache house. He was a favorite of Palache's two young daughters who were fascinated by his card tricks and his ability to pull an endless supply of 50-cent pieces from their ears. Jaggar also performed for adults. During a Thanksgiving dinner, he suddenly stood up from the table and feinted a cough. After covering his mouth with a large napkin, he proceeded to pull colored paper streamers from his mouth until they covered the entire table, much to the disapproving eye of Mrs. Palache. Another time, at Christmas, he dressed as Santa Claus and ran back and forth in front of the Palache house in the snow jingling bells so that the daughters could see him. He then climbed through a window with a pack on his back and gave presents to everyone. To Palache's daughters, he was "volatile and glamorous." And he caught the eye of Helen Kline.

She, too, came under the spell of the charming Harvard man. When they met, he had just returned from the Caribbean, tanned and fit. Though prematurely bald and with a large beaked nose and a crooked mouth—as he described himself—his overall features made him acceptably handsome.

And he was attracted to her. Helen Kline was young and wealthy and buxom. She was a woman of his own religious faith. And she reveled at social affairs. But Helen had her difficulties. She had a sharp tongue and had trouble controlling it. She was also impractical, buying overly elaborate hats and clothes that never lasted. But her biggest failure, as she freely admitted, was an inability to manage money, though she knew that she could write to her father "when what I had was spent."

Nevertheless, she would be the ideal wife for an ambitious man. And a courtship began—one that proved more difficult than most.

One evening during dinner a woman who Jaggar had known in Vienna arrived unannounced at the Palache house. The two Palache daughters were fascinated by her because the woman had arrived wearing a provocative dress, she smoked and she had arrived without a hat. After dinner, the woman and Jaggar went into a private room. The two girls raced to a secret place where they could hear anything happening inside the house. They heard the woman talk about marriage and money. In the end, the woman left. The two girls ran to tell their father what they had heard.

Charles Palache had seen such awkward situations before. And so he wrote to Helen's father, telling him of Jaggar's broken romances and dalliances in Europe. The father wrote to his daughter and demanded that she break off the relationship. But Helen was in love. Jaggar soon proposed. And she accepted.

Helen hurried back to San Francisco to prepare for a wedding, her father accepting her decision. He hired men to move furniture from the three main floors and put it in the attic or the basement or anywhere it could be stored. Heavy muslin was laid down to protect the carpets. Every room was repainted, given a different color scheme. Pink was the color of the master bedroom where presents would be displayed. Blue was the color of the main room where the ceremony would be performed.

A week before the wedding, Helen's mother hired caterers to begin working in the kitchen. Meanwhile, apparently oblivious to the activity all around her, Helen concerned herself with fittings for a gown and with ordering sets of frivolous hats. The Kline household was in an uproar when, in a rare display of anger, her father stopped all preparations and ordered his daughter to make do with what was on hand. He reminded her that, as a married woman, she would have to live within her means.

The wedding took place at the Kline house on the afternoon of April 15, 1903. Helen's only attendant was her sister, Eliza. Her brother, James, was the best man. The groom's father, Bishop Jaggar, performed the ceremony. More than a hundred people attended. Afterward, the newlywed couple boarded a train and headed back to Boston, the groom resuming his hectic schedule of teaching classes and writing scientific papers.

That fall, Jaggar was appointed an assistant professor at Harvard and the head of the geology department at the nearby Massachusetts Institute of Technology, the two institutions planning to merge their curricula. The next spring, he was elected to the American Academy of Arts and Sciences. He also found time to continue an acting career as an amateur, joining a pantomime group known as "The Strolling Players." In the summer of 1904, he led Harvard and MIT students on a field trip to the Black Hills of South Dakota. Helen spent most of the first two years of their marriage in San Francisco.

Still, Thomas Jaggar remained unsettled. In October 1904, he again applied for a position at the Royal British University in Melbourne, Australia. In the application letter, he gave the reason he wanted to leave Harvard and MIT.

> It is not for the salary, nor for discontent here, that I put myself in line as a candidate for the Melbourne chair. . . . It is for the reason that there is every prospect of my settling down here to too great a permanency, that I seize such a possibility as Australia. I have always preached travel to

    my students and would be untrue to myself if I did not
    try for every opportunity.

He was not selected for the job.

The next year, 1905, he organized a summer expedition to Iceland. The expedition would include himself, several other Harvard and MIT professors and several students. The expedition was to leave in May. They would travel to Scotland where they would board a ship that would take them to Iceland. Once there, travel would be on foot or on horseback.

A week before the planned departure, he excused himself from going, writing to the others that it was "impossible" for him to go because he intended "to spend the summer months in finishing my Martinique work and other matters." There was also another reason. Helen was three months pregnant. A son, Russell Kline Jaggar, was born at the end of summer on September 29, 1905.

Another academic year passed, Jaggar continuing to teach at Harvard and MIT. Then, on April 8, 1906, a newspaper headline again caught his eye. Vesuvius was exploding—and tens of thousands of people were fleeing for their lives.

Now Jaggar was determined. He was not going to miss an eruption of Vesuvius—which would prove to be the largest in almost 300 years—the way he had missed his own expedition to Iceland.

———•———

By April 8, Vesuvius had been exploding for more than a year. Admittedly, the explosions were minor and intermittent and, up to that time, few people considered them worrisome, and life continued as usual.

The first explosion occurred on May 27, 1905, when the people of Naples, who had a full view of the volcano, watched with amazement as a white cloud shot horizontally out from high on the volcano. Several minutes later, several streams of red lava poured slowly down the volcano in narrow streams. Within a day, the flow of lava ceased.

Similar explosions, also high on the volcano, continued for almost a year, spaced a month or more apart. And, again, there were a few short lava flows. None of this was enough to raise concern, even among those who lived on Vesuvius. Then, on the evening of April 7, 1906, a single deafening boom echoed across Naples Bay. Those who lived in the city, several miles away, looked toward the volcano and, this time, saw a single large cloud heavy with dark ash rise up quickly from the volcano's summit. More explosions followed. Each one sent out streaks of red-hot rocks that arched across the sky and fell on the lower slopes of the volcano. Then at least five streams of lava began to pour down the volcano, threatening everything in their paths. Those who lived on the volcano quickly packed their few possessions and, also carrying images of saints and the Madonna, praying all the while that the explosions would stop, ran for their lives.

A correspondent for *The New York Times* was in Naples at the time. He saw the explosions and watched lava cascade down the volcano and described the scene. "Torrents of liquid fire resembling distance serpents with flittering yellow and black scales, coursed in all directions, amid the rumblings, detonations and earth trembling, while a pall of sulfurous smoke that hovered over all made breathing difficult."

The next day, in a second report, the correspondent mentioned the unusual work being done by Raffaele Matteucci, the director of Vesuvius Observatory, who was charged by the Italian government with judging the danger. The correspondent ended his second report by saying "an American engineer named Perret was with Director Matteucci at the observatory."

———•———

Frank Perret once had a photograph of himself taken next to a steam vent on the floor of a volcanic crater near Naples. His slender figure is swallowed by an ill-fitting suit. In preparation for the photograph, he had placed the wide-end of a megaphone over the steam vent and had dropped a microphone, of his own design, attached to two wires, through the mouthpiece and down the vent. The other ends of the wires were connected to a headset, also of his own design. He is leaning forward concentrating on what he can hear. His straw hat has slid over his forehead, hiding part of his face. As he would tell others, this stance and the equipment he was using were the best way to listen "to the faint rumbles coming from the bowels of the earth." It is an odd photograph of an unusual man.

Born in Connecticut in 1867, an interest in volcanoes came early when, at age six, Perret watched his father, a Swiss immigrant and a seller and repairer of precision watches, hang a metal engraving on the wall of the family store in Brooklyn, New York. His father had accepted the engraving as payment for one of his watches.

The engraving showed the destruction of Pompeii by an eruption of Vesuvius. The boy spent hours studying the various features of the eruption depicted on the engraving. A great cauliflower-shaped cloud had risen from the volcano. Off to one side were smaller clouds crisscrossed with lightning bolts. At the bottom of the engraving were crowds of desperate people trying to get away and save themselves. Others had already died.

Ten years later, now age sixteen, two other events, totally unrelated, became guiding points in his life. The first occurred on May 24, 1883, and was the opening of the Brooklyn Bridge. From that night on, the glaring blue-white arc lights that outlined the bridge shone onto the tenement where the Perret family lived, illuminating the boy's bedroom. It was a constant reminder that electricity was the wave of the future—that its skillful application of new technology could change people's lives.

He was more awed by the second event, than inspired by it. It happened on the evening of August 30, 1883, when Perret was watching the setting sun. Instead of a normally blinding disk, he saw the sun "burnished like a copper ball." The subdued color was due to volcanic ash, shot high into the air four days earlier by Krakatoa. The ash was now circling the earth. That fact that a distant volcanic explosion could dim sunlight on the other side of the world intrigued the boy. The idea that powerful natural forces could be understood stayed with him the rest of his life.

Perret's formal education ended soon after these two events, at age eighteen, when he left the Brooklyn Collegiate and Polytechnic Institute—a preparatory school in engineering and science for boys who were planning to attend a university—and was hired by Thomas Edison to work in his laboratory in Brooklyn, not far from where the Perret brownstone was.

Edison assigned Perret to work with a team that was developing a safe and reliable lightweight battery. Perret had barely begun to work at the laboratory when workers went on strike. That caused Edison to close the Brooklyn laboratory and move it to New Jersey. Perret did not follow. Instead, he remained in Brooklyn, still living with his parents, and started his own company, the Elektron Manufacturing Company, which produced small, high-torque electric motors of his own design.

The young entrepreneur proved to be a poor businessman. After five years and few sales, he turned control of the company over to a banker, a Mr. John Barrett. Barrett reduced the product line from ten to two motors and moved the company to his hometown of Springfield, Massachusetts, a place better suited to manufacturing because it had lower rents than Brooklyn. It also had a history of heavy industry. Since the Revolutionary War, Springfield had been a major center for the manufacturing of ordnance.

There was an unexpected bonus to moving to Springfield. Its businessmen were interested in investing in new ideas. And Perret abounded in them. And so, in 1898, he started yet another company, the Perret Storage Battery Company.

The purpose of this second company was to produce lightweight batteries that could be used to power electric automobiles, a less noisy alternative to the gasoline-fueled contraptions that were starting to dominate the industry. In fact, Perret built several electric-powered vehicles. His most practical one was a 440-pound vehicle that could carry a single person at speeds of up to 13 miles per hour for a distance of 40 miles. The vehicle was never produced commercially. And the constant designing and building of vehicles—as well as the continued development of low-torque electric motors and lightweight batteries—took its toll on him.

In the spring of 1902, Perret suffered a complete breakdown of his health. While resting in bed, he read newspaper articles about the disaster at St. Pierre. As he later described his feelings, "The accounts of the complete annihilation of the city and its twenty-thousand inhabitants impressed me so much that volcanology, in which I had been interested from childhood, appeared to me to be the career for which I was destined."

And so, as soon as he got better and was able to travel, he sold his companies and headed for Vesuvius, one of the world's most famous volcanoes.

———•———

Perret arrived in Italy in late 1903. He took an apartment in the small town of Portici at the base of Vesuvius where the road to the summit began. He met local guides and hired them to take him on daylong hikes around the volcano. He took hundreds of photographs with one of the newly invented, folding pocket cameras, developing the photographs in a darkroom he had built in his apartment.

Nearly two years passed. Perret was in Naples when the first explosion of Vesuvius happened on May 27, 1905. He heard "a heavy detonation," then watched, as he remembered it, as "the long imprisoned lava burst forth in a cloud of steam and descended the mountain in splendid rivers of lava."

Sensing an opportunity, the next day he traveled up the road to the observatory and introduced himself to the director, Raffaele Matteucci.

Matteucci was known for his devotion to the volcano. Probably no one had ever been so enamored of Vesuvius and its eruptions. "I love the mountain," he was once quoted as saying. "She and I dwell together in solitude mysterious and terrible." Vesuvius, he freely admitted, was his mistress "whose wrath is more terrible than an army with banners." He vowed never to leave her. "I am wedded to her forever; my few friends say that her breath will scorch and wither my poor life one of these days." In fact, he had already been seriously injured. In 1900, while standing along at the edge of the summit crater taking photographs, a minor explosion rained hot rocks on him. He started to run, then stopped and turned to watch the action, in his words, courting "fate worse than that of Lot's wife." Falling rocks hit him, bruising his face and breaking his right leg. Somehow he managed to crawl down the mountain back to the observatory. He spent months recovering in a hospital, then returned and resumed his affair with the volcano.

After Perret introduced himself, Matteucci asked the American about his intentions. Perret answered that he knew how to build sensitive instruments that would record the activity happening unseen inside the volcano. He assured Matteucci that his intentions were noble ones. He wanted to understand volcanoes and be able to send out warnings to protect others. And Perret said he was willing to work for free. Matteucci accepted him, creating a new post for him, "Honorary Assistant to the Royal Observatory."

Perret moved into a room at the small hotel across the road from the observatory. It was sparsely furnished with two chairs, a table and an iron-framed bed. The legs of the bed had been embedded in concrete so that the bed would not slide across the floor during an earthquake.

Early one morning, almost a year after he moved into the room, Perret woke up and heard a low buzzing sound. He rose from his bed and leaned out an open window. All was quiet. He lay down again and the buzzing returned. After a brief investigation, he discovered the noise was coming up through his pillow. Using a technique he had learned from Edison, who, because he had lost most of his hearing, often clenched his teeth on a water pipe to determine whether his early phonographs were working properly, Perret got up and set his upper teeth against the iron bed frame. He felt a slight vibration. That morning he told Matteucci and suggested it meant an explosion would happen soon. Matteucci scoffed at the suggestion, answering back, "All you heard was the cook grinding coffee downstairs." A week later, Vesuvius did explode, the falling debris damaging a flower garden. Perret stood in the middle of the minor ruins with Matteucci and suggested that, perhaps, the cook was responsible for the destruction.

A few weeks later, on April 7, 1906, Perret was in Naples when he saw a "magnificent" dark cloud rise above Vesuvius. He took a few photographs, and then he hurried back to the observatory. He fought against a desperate crowd of people trying to get away. Some were carrying possessions. Others were clutching small statues of St. Anne, the guardian saint of those who lived around Vesuvius. He watched as some people stopped and prayed, asking the saint to stop the eruption. Part way back to the observatory, Perret paused and leaned against a wall. He felt a vibration so strong that it "moved the entire body to and fro like a pendulum."

Perret reached the observatory about midnight. Red-hot volcanic bombs were being hurled out of the volcano at a terrific rate, falling all around the observatory, bounding down the flanks of the volcano. Detonations increased in violence and in frequency. The volcano was pumping out a dust cloud at a prodigious rate, the cloud shot through with splendid electrical discharges. Each discharge had a snapping, spark-like quality. It was at this point in the eruption that Perret, though it was night, decided he would climb up to the edge of the summit crater, his way lit by the light of the eruption.

He walked almost a mile, but found he could only reach within several hundred feet of the crater's edge, the ground shaking too violently to allow him to approach closer. Here he turned and looked back down the volcano, surprised at how clear the air was, even though the giant eruptive column was shooting out of the top of the volcano right behind him. As he stood there, quite alone, he envisioned "the entire mountain as a huge boiler shell, humming and palpitating with internal pressure which was increasingly felt the higher one went." It was "a spectacle the eye could never tire of watching."

After an unknown length of time, he left and walked back to the observatory, amazed at how "effortlessly" the volcano had erupted. "Yet, the power was there to do it, and to do it easily and with dignity," he wrote later that night. "No word can describe the majesty of its unfolding, the utter absence of anything resembling effort, the all-sufficient poetry to do it majestically." He realized he was witnessing "one of nature's noblest spectacles; a symphony of violence." During his descent and return to the observatory, he stopped and looked back at the eruption. At that point, he raised his hand to his mouth and marveled at what he was seeing.

In addition to the explosions and the surge of material being thrown out from the volcano, there was electrical phenomenon happening all around him. On the observatory building, he saw electric sparks fire off around insulators and across lightning arrestors. The copper telegraph wires leading into the observatory were lit was a faint whitish glow. The eeriest effect, which Perret said he watched in dead silence, were the bright blue auroras that flickered in the air whenever a pocket of superheated air swept down and over the observatory. During such moments, anything metallic hisses and buzzes; everything, including Perret himself, was surrounded by a shimmering ethereal glow.

At three o'clock that morning, there was a tremendous shudder at the observatory as the eruption suddenly increased in violence as the top of the summit crater seemed to be ripping apart. Perret watched it through an observatory—one of the few windows not yet broken—describing the scene "as like the tearing away of the petals of a flower." The noise was like of Niagara Falls, only thousands of times louder. This was the critical moment and Matteucci ordered a retreat. He and Perret and four police officers who were still inside the observatory started down the volcano, having only rolled-up overcoats to put over their heads to protect them from falling rocks. At one point, Perret heard Matteucci remark, "This must be how Pliny died," a reference to the death of the Roman Pliny the Elder who died during the 79 A.D. eruption that destroyed Pompeii.

They could hear dull thuds all around them as rocks tossed from the volcano hit the ground. One of the police officers stopped to pick up a rock that had just fallen, dropping it because it was hot. He was then hit on the head by another flying rock.

The rain of stones lessened as the small group went farther down the volcano. Exhausted, they paused and made a fire. The air was unusually cold, thanks to the sea air from the Mediterranean that had been pulled inland by the draft caused by the furiously rising eruptive column. Perret, Matteucci and the four police officers stayed here for the remainder of the night, roasting eggs and eating onions, their only provisions.

All that night, the people of Naples were in a panic. Refugees from the eruption were crowding the city. Churches were open all night, the clergy doing their best to calm people's fears. A crowd gathered at the central telegraph station in Naples where Matteucci's hourly reports were being received. At three o'clock, when the eruption hit its most violent stage, the telegraph wire went silent. Everyone assumed the small party of scientists and police officer at the observatory must be dead.

But, the next morning, a courier brought news to Naples that the group at the observatory had survived. That evening, Perret sent a wire cable to his mother in Brooklyn, reassuring her that he was alive. Always frugal, Perret wrote: "Safe. In no danger. Frank."

———•———

Jaggar left Boston on April 10 for New York where he sailed for Italy, arriving in Naples on April 24. Two days earlier, Matteucci had declared the eruption over.

Jaggar made the trip to the observatory, riding the small electric train of the Vesuvian Railway. The cogged wheels of the train pushed slowly through drifts of volcanic ash. Trees along the way were bent far over, as if after a snowstorm, but these were covered with heavy ash. The gritty material was everywhere and seemed to penetrate everything. Any disturbance, no matter how slight, caused a swirl of the fine material to rise up into the air.

Just as Jaggar reached the observatory—he had to travel part way on foot—he saw two figures descending from the summit cone. They were Matteucci and Perret. Their faces and clothes were covered completely—as Jaggar remembered it, "picturesquely plastered"—by dust and ash.

After greeting each other, the three men entered the observatory building. One person who visited soon after the building was open in 1848 described it as "a Romanesque villa with the strength of a cannon-proof bunker." And so it was. The recent activity had formed a few new cracks in the thick cement walls and ceilings and almost every window had been broken, but the building was still structurally sound. Even the campanile-like outlook tower was still intact, as were the narrow terraces that wrapped around the second floor and provided external observation platforms.

Inside the building, Matteucci led Jaggar on a tour of the various rooms. One housed a library with bookshelves rising from floor to ceiling. In niches in the walls were paintings of mythical scenes. One showed the Roman goddess Minerva placing a crown on the head of Prometheus, the Roman god of fire. Another had Aeolus commanding the winds and yet another was of Vulcan, the namesake of all volcanoes, busy at his forge.

A separate room was a museum. Along the walls were shelves with bottles filled with ash. Each bottle was labeled with a date—the date that particular sample of ash had been erupted. In the center of the room were several enormous rocks set on tables. Each rock had a label that indicated when it had been erupted by Vesuvius. "These are very precious stones," Matteucci once told a reporter. "Some of them have hit me at one time or another. They represent my wounds."

The next morning, Jaggar made an ascent of Vesuvius, accompanied by three other visitors, all members of the Alpine Club of Great Britain. They were led by a local guide.

The wind was at their backs, which meant they were protected, at least, on the climb up, from having cinders hitting and stinging their faces. Nevertheless, the climb was a difficult one. The guide led them on a zigzag course, keeping to the occasional low ridge where the ground surface was compacted and staying away from gullies filled with loose sand. All the while, up and in front of them, steam boiled up silently from the volcano.

On the way up, they found the wrecked remains of the funicular railway that carried tourists—including Jaggar when he made his trip as a teenager in 1886—up to the summit. Only a few twisted rails could be identified. Everything else had either been blasted away or was buried beneath a new thick layer of ash and cinders.

After two hours of climbing, covering barely a mile, they stood at the crater's edge. "The fall-off was startling in the extreme," wrote Jaggar. The crater was so deep and the walls so steep that it looked as if the entire volcano had been reamed out. The guide cautioned the others that the edge formed an overhang in many places and might cave in.

The far side of the crater was obscured by smoke and steam. But, from the curvature he could see, Jaggar estimated is must now be about a half mile in diameter. Before the eruption, it was three hundred feet. He used an aneroid to measure the barometric pressure and, from that, determined that Vesuvius had lost four hundred feet of elevation, peeled away by the recent explosive activity. He walked close to the crater's edge, then lay down on his stomach and crawled to the very edge. He reached out and extended his hand as far as he could, touching the inner wall. It was cold and wet. He smelled the air. There was barely the odor of sulfur. No noise could be heard above the roar of a constant wind.

He backed away, stood and collected a few rocks from near the crater's edge. Then he and the others made their descent.

As they approached the observatory, they inspected themselves. Jaggar thought they looked "ludicrous to the extreme." Their eyelids, noses, ears and hair were heavily caked with gray ash. It was impossible to identify the color of their clothes. And his new camera, a collapsible Kodak Brownie, was no longer new.

Jaggar stayed at the observatory four more days, Perret showing him the collection of photographs he had taken of the eruption and the crude devices he had used to record activity. The simplest was a small saucer filled with liquid mercury. Any slight vibration set the heavy liquid to wriggling. Perret had also hung four simple pendulums from a horizontal post. Each pendulum was of a different length, and so swung at a different natural frequency. By watching pendulums swing, Perret concluded that most of the earthquakes coming from Vesuvius shook the ground back and forth at a frequency of about one second. He also showed Jaggar a notebook in which he had kept a record of the intensity of earthquake shaking. He had made the determinations by standing with his back against a cement wall and counting how many times he was pushed back and forth. But the most remarkable record was his eyewitness account of the eruption—the eerie electrical phenomena, the fantastic eruptive column.

"I knew at once," wrote Jaggar of their meeting, "that Perret was the world's greatest volcanologist." He saw in him a kindred spirit. But there was an important difference between the two men. In his study of volcanoes, Perret was not constrained by the demands of a regular job. It was the life that Jaggar was seeking.

———•———

Three weeks later, Jaggar was back at Harvard. A letter was waiting for him from Gilbert Grosvenor, the president of the National Geographic Society. In the letter, Grosvenor said that an article was being prepared for the Society's magazine about the 1906 San Francisco earthquake, which had occurred on April 18, while Jaggar was en route to Italy. Grosvenor was writing to ask whether the Harvard professor might write something for the magazine about Vesuvius and its recent eruption.

The pair of articles about the San Francisco earthquake and the eruption of Vesuvius appeared in the July 1906 issue of *National Geographic* magazine. Jaggar might have expected praise from his Harvard colleagues for having published a report about his trip so soon, but the man who had funded his trip to Italy was furious.

Alexander Agassiz was an introverted and dour man who could be gruff and, to most, was a severely intimidating figure. By contrast, his father, Louis Agassiz, one of the great naturalists of the age, was charismatic and robust. He was often animated and always seemed to be good-natured. It was Louis Agassiz who first championed the idea that great sheets of ice had recently covered much of North America and Europe and thereby dramatically altered our view of earth history. As a Harvard professor, Louis Agassiz had founded the Museum of Comparative Zoology and was its first director. After his death in 1873, the directorship passed naturally to his son, Alexander, also a Harvard professor, who not only provided capable leadership and was a well-respected naturalist in his own right, but also contributed much of his personal fortune to maintaining and running the museum.

Unlike his father, for whom the acquisition of money always eluded him, the son amassed a great deal of personal wealth. It came after he acquired the ownership to two failing copper mines— the Calumet and Hecla mines—in upper Michigan. He turned the mines into highly successful businesses, improving the efficiency of the extraction process and expanding the transportation system. For decades, the Calumet and Hecla mines were ranking among the world's most productive mines and produced more than half of the nation's copper. It was with these profits that Alexander Agassiz greatly expanded the museum at Harvard. In 1901, he added a geologic section and began to fund geologic expeditions, including Jaggar's trip to Vesuvius in 1906. Today Alexander Agassiz is remembered for his great generosity and for his own professional contributions, especially his exploration of coral reefs. For seven years he served as president of the National Academy of Sciences. Today the Academy bestows a prestigious medal for research in oceanography in his name. Alexander Agassiz is also remembered for a personal trait: He was known to be vindictive.

One of the first to suffer was a promising young zoologist named Jesse Walter Fewkes. Fewkes, a native of Massachusetts and a Harvard graduate, had worked as an assistant at the museum for more than a decade when he decided that his career was going nowhere and decided to explore other options. He made a trip to California where, so it was rumored, though the rumor was never substanti-ated, he was offered employment. That alone was enough to raise the wrath of Alexander Agassiz who informed Fewkes that his association with the museum was over. Later, in a letter, Agassiz would condemn Fewkes as "a consummate sneak and hypocrite." Another assistant was Charles Otis Whitman who had worked at the museum for twenty years and was forty years of age when he was summarily dismissed from the museum for his supposed involvement in teaching at another museum. Jaggar, too, was to run afoul of Agassiz.

In Agassiz's eyes, Jaggar's sin was that he had written an article about his trip to Vesuvius for the National Geographic Society without first informing Agassiz. Agassiz wrote to Jaggar, saying that Jaggar's underhanded deed "smacked of self-promotion." Agassiz said that he "disliked this kind of advertising." He then informed the much younger Jaggar that "if you wanted to publish a prelimi-nary notice the Museum Bulletin was accessible to you."

Jaggar responded in the way most people would when confronted by someone who was senior to them. He apologized.

> I sincerely regret that the narrative in the National Geo-graphic magazine was published, in view of what you tell me. . . . I felt that I could not refuse to send something, because the National Geographic Society sent me money, specially appropriated, while I was in Martinique and I have never done anything for their magazine.

But that was not enough to placate Agassiz. Less than a month later, two articles published in Boston newspapers about the Vesuvius eruption contained quotes from an unnamed Harvard professor. Again Agassiz wrote to Jaggar accusing him of being the unnamed source.

> I saw the other day, to my great regret, one other of those reportorial articles . . . to which I have called your

attention. From internal evidence and quotation marks it is very evident that the foundation of this notice has been supplied by you.

Agassiz continued by lecturing the assistant professor on the unseemly use of the popular press to report scientific results.

There are two methods by which a man's work may be known that of the modest investigator known to his colleagues and peers outside of the institution with which he is connected, and whose name never appears in the daily press. The other is the method which you seem to be devoted to of obtaining through the papers an ephemeral and cheap notoriety which will never bring you the recognition of your scientific colleagues.

Agassiz concluded the letter by writing:

I am not in [a] position to waste my time . . . herewith notify you that I cancel all my agreements with you relating to publications of Mt. Pelee and Mt. Vesuvius and waive all claims I may have to publishing your materials. . . . I also withdraw the nomination I intended to make to the Museum faculty to allow you to study the volcanic rocks I have collected at various times in the Pacific. These rocks will remain in my possession as they are my personal property.

Yours very truly A. Agassiz

This time, instead of responding directly to Agassiz, Jaggar forwarded a copy of Agassiz's letter to Harvard President Charles Eliot and added a note of his own. It informed President Eliot that he was resigning his professorship at Harvard, effective September 9, 1906.

CHAPTER SIX

ALASKA

J aggar's sudden resignation from Harvard was a mixed blessing. It eliminated a time-consuming commute to teach classes at Harvard in Cambridge and at MIT in Boston. It also reduced his income. With a wife and a son to support, he realized he needed to earn additional money. Yet, he obviously wanted to continue a pursuit of volcanoes and their eruptions. He found a way to do both.

Alaska was in the minds of many Americans during the first years of the 20th century, in part, because of the recent Yukon gold rush, and because much of Alaska was still unexplored. And so, Jaggar reasoned, as did others, unclaimed mineral wealth must still lie in that vast country. Alaska also has one of the longest chains of volcanoes in the world—more than sixty volcanoes that stretch for two thousand miles from the western tip of the Alaskan Peninsula to Kamchatka at the eastern edge of Asia. These are the Aleutian Islands. And so he organized an expedition to search for mineral deposits on these islands—and to explore the volcanoes.

In a letter soliciting funds from investors who might be interested in financing such an expedition, Jaggar quoted a recent report from the United States government that suggested both mineral ores and coal might exist on at least two Aleutian Islands, Unalaska and Attu. According to the report, on these two islands, there exists "green-stones, slates and granites of the same age and kind as those which bear copper near Prince William Sound." Jaggar guaranteed that each investor would share in any claims he might file, the amount of a share to be proportional to the size of an investment. As the expedition's leader and organizer, he would retain "10% interest in any property developed."

To his surprise, he quickly raised the needed money. Years later, he reflected on what he learned from the experience.

> The great revelation of my life I got [in 1907] from the rich copper brokers of the Boston stock market. An MIT official told me to go down on State Street, organize in writing the Technology Expedition to the Aleutian Islands, tell them I wanted ten thousand dollars, that it was mere exploration, that there might be copper or there might not, and that I was going to start in April. He gave me a list of big copper men. There happened to be a scientist Alexander Agassiz at the head of Calumet and Hecla, and he disliked me at the moment. I went into that office with fear and trembling and in three minutes came out with one thousand dollars as a first subscription. In one week I raised $13,000 to my utter astonishment. I made mimeograph reports about every two weeks, told them everything bad or good, bought a schooner, took student assistants, made lots of mistakes, muddled through, struck a huge market depression in the autumn, and published in the Technology Review an

adventurous account of the trip which contained precious little geology and a whole lot of Bull.

With the money in hand, Jaggar next identified his scientific team. He enlisted Arthur Eakle, a mineralogist from the University of California, and Harvard astronomer Henry Gummeré who, years earlier, had helped Jaggar build a water tank to produce ripple marks. Two senior students in mining engineering from MIT, Desaix Myers and Harry Sweeny, joined. To ensure the health of the men, Jaggar included a physician, Dr. Edwin Van Dyke of San Francisco, who would also serve as the expedition's entomologist and botanist.

The team assembled in Seattle on April 22, 1907, at the Hotel Lincoln, Seattle's most luxurious hotel, Jaggar apparently not sparing any expense. He set upon examining all the ships that were for sale in and around Seattle. After almost a week of searching, he settled on a thirty-nine-ton schooner, the *Lydia*, a sturdy vessel built twenty years earlier for the north Pacific sea-otter trade. He purchased it and all accoutrements—extra sails, cordage, food and medical supplies, fishing gear, bedding, charts, navigational instruments and two fishermen dories, the last to be used to land men in harbors and to explore uncharted bays and islands—for $4,000, almost a third of the expedition's funds. The man who sold him the *Lydia*—a Mr. Carl Guntert—had, at first, agreed to serve as captain, but, within a week of the ship's purchase, he announced his determination not to go, and so a delay followed as Jaggar searched for another captain.

Jaggar was eventually introduced to George Seeley, a Nova Scotian, who, for the last five years, had captained a small sailing vessel that carried sugar from the Hawaiian Islands to San Francisco. But Seeley had a cloud over him.

Six years earlier, in 1901, he had been captain of the steamship *Oregon* when it was carrying three hundred passengers from Nome, Alaska, to Seattle. Three days out, control of the rudder was lost and the *Oregon* drifted for almost a week. Seeley put the passengers and the crew on short rations—coffee and hardtack for breakfast and an abbreviated meal served in the afternoon. There were the anticipated strong complaints. Fortunately, the captain of a sister ship, the *Empress*, spotted the *Oregon* and towed her to a port. An investigation was conducted and Seeley was cleared of negligence with regard to the rudder, though questions were raised about his treatment of passengers and crew. Ship owners in Seattle were worried that Seeley was at the beginning of a streak of bad luck, and so no one in that city had hired him to captain a ship to Alaska since.

Jaggar did not have a choice. It was mid-spring and those ship captains who were familiar with the treacherous waters and the frequent rough weather of Alaska had either departed already for a summer of sailing or had signed contracts with other ships. And so Jaggar hired Seeley.

There was a bad omen just before Jaggar planned to sail. Seeley had assembled a crew consisting of a mate, a cook and five seamen. On the night before the *Lydia* was to sail, two of the seamen slipped off the *Lydia*, taking their possessions with them. Seeley and Jaggar spent the next morning searching the Seattle waterfront for the missing men, who had already been paid half-wages, but neither man could be found. It was May 20, a late date for a departure to Alaska, and so Jaggar had Seeley call off the search.

Late that afternoon, the *Lydia* was cast off from a buoy and towed a mile by a gasoline launch until it was clear of shipping. Then, as soon as the sails were set and Jaggar hoisted the MIT pennant, the wind dropped, another bad omen. The crew and the party of scientific men spent their first night on the *Lydia* bobbing in a calm sea a few miles from Seattle, still within sight of the city lights.

———•———

At sunrise a slight wind came up and the sails were set again. Seeley showed his sailing prowess by taking the *Lydia* more than a hundred miles the first day to Protection Island at the eastern end of the Strait of Juan de Fuca between Washington State and Vancouver Island. By the morning of the third day the *Lydia* had covered more than two hundred miles. As planned, Jaggar had the schooner sail into the harbor at Port Angeles where he and his colleagues would spend the next four days on land trying out and reorganizing their camping equipment. (Jaggar had bought the equipment at Abercrombie & Fitch in New York. As at the hotel, Jaggar did not spare expenses.) On May they were back on the *Lydia*. The next night they were far out at sea.

Jaggar's original plan was to wind his way through the islands of the Inside Passage that form the west coast of Canada and include the southeastern part of Alaska. From there, he intended to make a straight crossing to Attu Island at the extreme far western end of the Aleutian Islands where the formal work of the expedition would begin. From there, he could take advantage of the prevailing westerly winds to return, stopping and exploring the volcanic islands as he went. But the late start—and the fact that the *Lydia* did not have auxiliary power—forced him to change the plan. Instead, he directed Captain Seeley to sail straight across the North Pacific to Unimak Island at the eastern end of the Aleutian Islands. From there, they would sail west and visit as many islands as time and as the weather allowed.

The passage to Unimak Pass took twenty-nine days, mostly beneath a cloudy leaden sky. Throughout that time, as Jaggar noted, the ship was followed by a "barnyard" of birds, which included gooney and albatrosses. A giant kelp, forty-three-feet long, was picked up on June 6, covered with barnacles, small crustaceans and algae. Whales and porpoises were seen from time to time. One day, a fur seal was spotted far out in the Pacific. The scientific team spent much of their time overhauling equipment, preparing food in sacks for the land trips, reading, and helping the crew to scrape and paint whatever needed to be scraped and painted. On June 22, during a hard blow from the north, the *Lydia* proved itself tight in weather, but those unfamiliar with the sea were caught by surprise. Once, when the ship heeled far over, Jaggar was in the galley and was caught unaware on the upslope side of the ship. Trying to manage both a cup of tea and a plate of dinner, he lost his grip and slid across the cabin deck into dishes, wrecking his plate and pouring hot tea down his neck. As he later judged the incident, "This was a common diversion."

On June 27, at five o'clock in the morning, land was sighted. Seeley's navigation had proven accurate. It was Unimak Island. The weather was fine, and the wind light. That day, the cook caught two codfish. There were many gulls, puffins and "whale birds" around. Just after noon, the *Lydia* tacked up against a northwest wind. The snow-covered heights of Pogromnoi volcano came into view.

Unfortunately, sea conditions were too rough to permit a landing with the dories that the *Lydia* carried, and so Jaggar instructed Captain Seeley to bypass the island and the volcano and to sail west. The same happened a few days later at the paired islands of Akun and Akutan. Both have volcanoes with steep sides. On Akutan, one of volcanoes was throwing up columns of black smoke. Jaggar estimated the height of the volcano to be four thousand feet, the same as Mount Pelée and Vesuvius. But, again, a landing was impossible. This difficulty in landing would be the great frustration of the trip—to come within a mile or so of an unexplored and steaming volcano and being unable to land and have to sail on. The *Lydia* finally made its first landfall at Dutch Harbor on Unalaska Island. Here Jaggar hired a guide to lead him and the others fifteen miles overland to Makushin volcano.

In preparation for the trek, he had the men carefully weigh and apportion the food, then, with tents and kits, they took off, each man carrying about forty pounds in a rucksack.

They marched all day in the rain, covering several miles. That night they camped on a grassy terrace. Dinner was bacon and flapjacks. The rain continued all night and all the next morning. In fact, it rained so often that Jaggar wrote in a journal that any future reader should assume it was raining unless he specifically said that it was not.

On the third day, the men started climbing the steep, snow-covered slope of Makushin. "The climb was long and easy," Jaggar wrote. "The crevasses are few and small." On the way to the top, they passed several sulfur-crusted patches on top of new snow. In places, they had to detour around deep holes, seeing black mud bubbling at the bottom of each one. At the summit, they discovered a large steaming crater, a feature unknown to the guide. Jaggar named the new feature "Technology Crater."

On the way back to Dutch Harbor, Jaggar separated from the others so that he could climb a rock face partly covered with ice and snow. He soon found himself stranded on a ledge barely wide enough to stand. His presence startled two eagles that took flight. They soared around him in great circles, then out over the nearby ocean where Jaggar could see whales sounding, their enormous forked tails rising straight up, then slowly sinking.

The two eagles soon came back. They circled him again, moving their heads from side to side. "They know I am helpless," Jaggar thought.

It then rose in his mind that he might have to stay plastered to the rock face hundreds of feet above the sea until help came. He had an idea. He dropped his pack and watched as it "bounded suggestively from ledge to ledge and rolled in fragments to the water's edge below." Then he jumped.

As it turned out, the downward slide in a leather coat, heavy boots and gloves was not a difficult task. And it gave him a chance to look back.

"Behold," he wrote of the minor adventure, "there was the eaglet solemnly cuddled into a corner of the cliff, and all the nature-study stories came back to me with startling vividness."

This was the world he had dreamed about during his childhood. Here he was surrounded by wilderness. And that reminded him why he had decided to study geology—because it offered such varied and unexpected adventures in a rapidly modernizing world.

The better part of the day over, he marched back to Dutch Harbor, now "wiser concerning the peculiarly high grades of Aleutian slopes, masked by long grass." The weather was calm. He rejoined the others. He spent the evening watching "forty or fifty whales making the bay breathe." Then he wrapped himself in a sleeping bag and went to sleep.

The next day, back at sea, Jaggar asked Captain Seeley to make for Bogoslof Island. For two days, the *Lydia* beat against a strong headwind. The ship making little progress, Jaggar took a measure of the situation, of the price exacted on the men after six weeks at sea. "The mate is prostrate with a strained back, the captain has boils and a wounded hand, and nearly everybody in the cabin is sea-sick." And so he decided to turn the ship and head back to Dutch Harbor where they would wait for better weather.

A week later, on July 14, the government revenue cutter *McCulloch*, which had the dual task of carrying the mail and chasing off foreign ships that were hunting seals in American waters, sailed in Dutch Harbor, having completed a run to faraway Attu. Jaggar talked to the crew. They had passed Bogoslof two days earlier and had seen a new steaming cone. Excited by the report, even though the weather had not yet improved, Jaggar and Seeley set sail immediately. This time the *Lydia* came within sight of the small barren island, which was steaming vigorously, but the wind was too strong to attempt a landing, and so the volcano was bypassed and the *Lydia* continued sailing west.

After three more days of sailing, the *Lydia* arrived at Umnak Island. Here was a sandy beach with a gentle slope where a landing could have been made, but the wind was strong and the surf was high. And so, yet again, the *Lydia* and its men turned away. Jaggar, frustrated by so many missed opportunities, watched as yet another unexplored volcanic island faded into the distance.

After another four days, sailing under gradually clearing skies, the *Lydia* arrived at the islands of Amlia and Atka. Amlia showed a long, rugged, though not very high profile. Atka was mountainous, rising from the sea as several large glaciated peaks. Each one was a volcano. The highest was Korovinski on the north side of the island. After some delay because of a lack of wind, the *Lydia* finally sailed into a cove where there was a village of sod huts and frame houses that contained the whole population of Atka, one hundred and ten souls.

The scientific party landed and stayed on Atka for ten days. The physician Van Dyke collected plants and netted insects. The mineralogist Eakle and one of the students, Myers, worked on a crude geologic map of the area around the cove, noting the locations of volcanic cinders and of warm water. The astronomer Gummeré set up equipment to measure the direction and intensity of the magnetic field. Jaggar and the other student, Sweeny, packed their rucksacks and set out to climb Korovinski.

For three days, the pair circled the base of the volcano, looking for a way to the top. The rain and the cold wind were relentless. On the fourth day, they found a hot pool where steam was rising from the rocks. Here they spent a day and rested and ate their meals in the steam bath. "The warmth was very grateful in contrast to the cold wetness of everything else, for we were already soaked to the skin," Jaggar recorded in his journal.

On the fifth day, still searching for a way to the summit, the constant rain turned to a dense fog. At midday, Jaggar and Sweeny set up their silk tent and crawled into their sleeping bags, hoping the weather would change. Then the wind came up. For the next two days, they were stormbound, taking turns holding up the center tent pole, even at night. Occasionally, a blast of wind caused one of the tent pegs to pull out of the ground and the sides of the tent to flap. Then one of them had to go out, get wet, and drive the peg in, then pile stones on it.

During the second night, the storm increased. The food was down to some hardtack and dried beef. Both were too tough to chew and too dry to swallow. "It would be difficult to devise a more completely miserable situation," Jaggar wrote of that night. Here they lay, "baffled, wet, hungry, midway between America and Asia, between the Arctic and the Pacific, between earth and sky, and far from any warmth or any base of supplies."

Fortunately, the third day began with gray daylight, though the rain continued. The storm had wrecked the tent, and so Jaggar and Sweeny left it behind and packed the rest of their equipment and marched toward the coast, plunging down steep slopes amid sheets of driving rain and over a plateau of wet and slippery grass. Where possible, they used the wet grass to their advantage to speed their descent to the ocean, each man using his rucksack as a toboggan, careening down slopes, thankful when they were finally in sight of the *Lydia*—and that they had no broken bones.

Back on the *Lydia*, the cook fixed them hot food. That night, they "turned into our beds and slept the sleep of the blissful."

Atka is midway along the chain of the Aleutian Islands. It was now midsummer, and Jaggar realized it was time to start the return. Before he left Atka, he gave the islanders whatever cheese they had left and a sack of flour. In return, the islanders handed him several baskets woven from native grass and made by local women.

The *Lydia* left Atka on August 4. Three days later, Bogoslof was again in sight. The rocky crags were still steaming. Jaggar instructed everyone to prepare "for a vigorous investigation" of the island.

———•———

The first recorded sighting of the island was in 1769 by the Russian explorers Mikhail Levashev and Peter Krenitzin. Years later, they published a nautical chart that showed the location of the island, depicting it as four small crosses to indicate the presence of precarious rocks.

The next sighting was by Captain James Cook during his third and final voyage of the Pacific Ocean. He sailed close to the island on October 29, 1778, describing it as "an elevated rock like a tower." He did not give the island a name.

On May 8, 1796, people on Umnak Island, thirty miles to the south, saw a column of smoke rising from the sea. Then darkness descended over the island caused by a rain of volcanic ash. That night, after the fall of ash had stopped, they could see a bright red light illuminated on distant clouds and they could hear a terrific roar. Those who went to investigate found a new rocky island about a half-mile north of the one seen by Levashev and Krenitzin, and by Cook. It was Russian traders on Umnak who named the new island *Joánna Bogoslova*, John the Theologian, because the eruption had started on the feast day of Saint John. Today any land that appears in that part of the chain of Aleutian Islands is known as Bogoslof.

The eruption that began in 1796 continued for several years. The first people to record an actual landing on the island were sea-lion hunters in 1804. By then, the island was about three miles in circumference and five hundred feet at its highest point. The hunters reported the ground was so warm that it was difficult to walk on it.

What happened during the next several decades is not well documented, but, in the summer of 1883, captains of passing ships noted that the island was constantly obscured by a combination of steam and ash. Finally, in September, the steam and air cleared and ships were able to sail close to the island. The volcano was still erupting, throwing out masses of red-hot rocks and volumes of ash and steam. A second island, New Bogoslof, was now spotted. This activity ended sometime before the end of the 19th century, the second island separated from the first by a channel about a mile wide.

Activity resumed in March 1906. By September, someone on a passing ship managed to take a photograph of the island. It showed both New Bogoslof and Old Bogoslof were rocky crags. Midway between them was a large steaming mound.

Jaggar carried a copy of the 1906 photograph as he prepared to land on Bogoslof. Even from the deck of the *Lydia*, he could see that dramatic changes had occurred in less than a year. Now two mounds occupied the central part of the island, the newer mound, according to Jaggar, "steaming like a pudding."

He and his scientific team made their landing on the morning of August 7, 1907. Hundreds of sea lions, bellowing with voices that justified their name, swam within a stone's throw of the two dories as the men rowed toward the island. Occasionally, one of the animals would raise itself high from the water to stare, then plunge frantically beneath the waves. Once the men reached land, the sea lions that were on shore hurried into the water, except one immense bull that might have been asleep. Eventually, he, too, headed for the sea, slowly, giving the men ample time to take photographs of him before he slid into the water.

Each man hurried off to take up his own specialty, knowing he had only a few hours on the island. Jaggar began by drawing an outline of the island, which was now two miles long at its greatest extent. Jaggar also sketched a crude topographic map. Next he measured the dimensions of the two central mounds, determining that each was about two thousand feet across at the base and stood about four hundred feet high. Between them was a crescent-shaped lagoon. Jaggar walked to the edge. The rocks underwater were stained a bright orange. The surface of the water was steaming slightly. He bent down and tasted it. The water was salty. He took out a thermometer. The water temperature was a pleasant 90°F. He then set off to circumnavigate the island on foot.

At the southern shoreline he found a long rocky terrace. At one end of the terrace was a cave. Both the terrace and the base of the cave were twenty-five feet above sea level. Both had clearly been made by the action of breaking waves. He could identify both on the 1906 photograph he carried, which showed both at sea level. The conclusion was undeniable: In less than a year, this part of the island had risen twenty-five feet.

He found evidence of recent uplift elsewhere on the island. There were more terraces, raised sandy beaches and notches on rock outcrops that had clearly been cut by sea waves. All were high above present sea level. Had the island been lifted up in one great thrust or had the movement been gradual? Was it rising now as he stood there? Was only the island rising, or was the entire sea floor that surrounded it, carrying Bogoslof and the other Aleutian Islands upward? And where else in the world might such activity be happening?

To answer such questions, as Jaggar now knew—he had been thinking about it since St. Pierre—required a new type of scientific institution, a *geonomical* observatory—his term. They would be based on the volcano observatory at Vesuvius, in that they would have a permanent staff of scientists who would be using the best equipment to measure earthquakes, collect gases and record slow crustal movements such as he was seeing at Bogoslof. There would be hundreds of such geonomical observatories to study the earth's internal activity, in much the same way that hundreds of astronomical observatories already existed to study the stars, and they would be located where the earth's activity was greatest, that is, in Alaska and along the Aleutian Islands, in Japan, in Italy, along the coasts of North and South America and throughout the Caribbean.

"The cry of suffering multitudes, which led Pasteur on from ferments and silk-worm diseases to hydrophobia," Jaggar would write at the end of his official report of his expedition to the Aleutians, "was no more heart-rending than the cry of the terrorized millions who live in earthquake and volcano lands." To alleviate "the cry," a worldwide network of geonomical observatories was needed to accumulate critical knowledge about "this old earth, which is pushing up and down its shore lines in a hundred places not yet explored" and where the earth was "building other Bogoslofs."

———•———

After leaving Bogoslof, the *Lydia* stopped again at Unalaska to replenish its stores at Dutch Harbor, then sailed east to Akutan Island, the expedition's last stop. The scientific team landed, but, as would be a persistent theme in this expedition, a terrific storm kept them from reaching the island's volcano. Instead, they settled for a few days of fishing for salmon in a nearby river, using their knives to stab the fish, sometimes catching a salmon with their own hands and throwing the fish up onto the bank.

On August 22, as the storm continued, the crew of the *Lydia* prepared the ship for the long trip back to Seattle. That night, an earthquake shook the island, a reminder of the geologic activity of the Aleutian Islands. The next morning, the *Lydia* headed home, an uneventful passage of eighteen days of fair weather and frequent strong winds, the ship sailing along a straight course.

Soon after he arrived in Seattle, Jaggar learned that, while he was midway across the North Pacific—to be more precise, twenty-four days after he and his colleagues had stood on and examined

Bogoslof island—members of the crew of the Pacific whaler *Herman*, sailing south from the Bering Sea to the Pacific Ocean, were "spellbound" when they passed Bogoslof. A dense black cloud was rising from the island, and the air was filled with the smell of sulfur. The sea for miles around was churning and mixed with "volcanic earth." From the deck of the *Herman*, the crew watched as steam shot up, at irregular intervals, from "the ocean's vital parts." Several of the crew urged the captain to sail closer, but he refused. By then, ash and sand were falling at Dutch Harbor, forty miles southeast of the eruption, where people were also hearing the clamor of distant thunder.

On October 15, the cutter *McCulloch* sailed close to Bogoslof. As the ship approached, the captain recorded in the log that "there was but little steam escaping" from the island. He was unprepared, however, for the changes that had happened since he had seen the island the previous summer. As he recorded his initial view, "the first change to be observed was at such variance with anything that had been expected that it was startling, to say the least."

As the *McCulloch* sailed closer, it was unmistakable. The two central mounds were gone. In their place was a new steaming lagoon, a half-mile in diameter, and the water of the lagoon "in a constant state of ebullition." Elsewhere, the formerly rugged outline had been "softened by a padding of lava dust that almost disguised the island beyond all recognition." The captain of the *McCulloch* estimated that "incalculable tons of material, hundreds of feet in depth, had been deposited over the entire island."

In his official report about the expedition, Jaggar wrote that he regarded the hours spent on Bogoslof as "the most interesting of the whole trip," describing the volcano to a newspaper reporter, after he returned to Boston, as the "island that changed form while you wait."

———•———

The voyage had been difficult, but deeply rewarding. "Lively the place was in every sense," he wrote of the experience, "the hot earth alive, heaving and heaping, the sea alive, currents, surf and warm lagoons; the shore alive with the hundreds of immense clumsy leviathans, bulls, cows, pups; and, finally, the cliffs alive with their teeming bird-life." But, now, it was time for him to return to his life in Boston.

Jaggar resumed his teaching and administrative work at MIT. A few weeks before Christmas, he received three blue fox skins, which he had bought from a trader on Unalaska and which he would give to his wife as presents, he remarking that the skins turning "out most successfully after treatment by an eastern furrier." Also, just before Christmas, he heard that President Theodore Roosevelt was sending a battle fleet of the United States Navy on a cruise to ports in South America, a cruise that would eventually take the fleet around the world. Jaggar wrote directly to the President, suggesting that a geologist be included as one of the crew. To no surprise, he volunteered himself.

"My suggestion is based on the belief that the opportunity is a rare one," he wrote to Roosevelt, "in which to see comparatively—in a brief space of time—many seismo-volcanic shores and to examine the seismometrical methods of different governments and organizations." In the letter, he noted that he had succeeded in getting a resolution passed at a general meeting of the Geological Society of America that "strongly recommends to the several North American governments and to private enterprises the establishment of volcano and earthquake observatories." He reminded the President of the recent destructive earthquake in San Francisco and the fatal volcanic eruptions at Martinique and in Italy and that Panama, where Roosevelt had decided to build a canal, was a region prone to earthquakes, as was the Philippines, where there were also numerous volcanoes, a country possessed by the United States since the end of the Spanish-American War. Sending Jaggar with the fleet of Navy ships—which the press had dubbed "The Great White Fleet"—would be a step toward establishing the volcano and earthquake observatories recommended by the Geological Society of America and bear directly on the prediction of such natural events and on the protection of life and property.

The Secretary of the Navy responded for the President. He wrote to Jaggar, saying "that it would be impractical to have a geologist attached to the battleship fleet; also that the itinerary of the route was such that very little if anything could be gained by a geologist."

And so, yet, again, another opportunity passed.

———•———

In May 1908, his mother, Anna Lawrence Jaggar, became gravely ill. Hoping to improve her health, Bishop Jaggar accepted the mostly honorary position as head of American Churches of Europe, and the two of them moved to southern France where the climate is good. On August 31, Anna Jaggar died.

Her father had been bequeathed a trust of $25,000 to her with the stipulation that the interest on it would be paid to his daughter as long as she lived, then, after her death, the principal would be divided equally among her surviving children. Jaggar and his sister each received a share. He already had plans on how to spend the money.

Even before his mother's death, Jaggar was planning to return to Alaska the next summer. The expedition would be bigger than the first. Both the United States Geological Survey and the Imperial Geographic Society of St. Petersburg in Russia had agreed to contribute to the second expedition. Now, with an inheritance, he envisioned a grander scheme.

As he saw it, a return to Alaska would be the beginning of "a series of scientific expeditions that, eventually, would cover all of the volcanic regions of the world." The work would culminate, after a decade of travel, with the publication of a multi-volume work that would recount his many anticipated adventures and include hundreds of photographs to illustrate his work. He estimated that the entire project—the exploration and the writing—would require at least twenty years, the remainder of his professional career. To encourage contributors, he wrote a letter, saying that he was prepared to resign from MIT if sufficient money was raised. To indicate his personal commitment to the project, he also wrote, "I have twelve thousand dollars pledged toward this work in one anonymous gift from my own family." His wife, Helen, disliked the plan.

Ever since their marriage, they had shared a house with another family. Helen realized that her husband's inheritance was probably the only one he would ever receive. Furthermore, he was planning to leave soon on another trip to Alaska. The first one had lasted four months. Additional ones would probably last longer. And they had the future of a young son to consider.

After what was probably much intense discussion, they came to a compromise. They would use the inheritance to purchase house. The remainder would be used to pay for a trip for both of them.

Japan was then much in the news, having recently opened its borders to foreign travelers. Also President Roosevelt had just mediated an end to the Russo-Japanese War for which he had received the Nobel Peace Prize. Those two events had sparked a brief craze in the United States for anything Japanese. Magazines often had cover articles that described something about the mysterious and exotic country. American clothes designers were incorporating Japanese motifs in their creations. The most popular wallpaper showed Japanese-inspired designs. Society parties often had Japanese themes with pseudo-tea ceremonies. All things Japanese—Japanese prints and porcelain, judo and Buddhism, geisha and samurai—were suddenly in vogue. And, as Jaggar knew, Japan was also a land of volcanoes and of frequent earthquakes.

And so he and Helen agreed. They would go on a summer trip to Japan. And, on the way, they would take advantage of a stopover in the Hawaiian Islands to see the remarkable lava lake at Kilauea volcano.

CHAPTER SEVEN

THE PACIFIC WORLD

The Jaggars sailed from San Francisco on March 26 on board the SS *Siberia* of the Pacific Mail and Steamship Company, "the most Luxurious way," according to a company brochure, "to travel the Sunshine Belt to the Orient." They traveled as first-class passengers, as did more than two hundred others. One of the others was Queen Lili'uokalani, the last reigning monarch of the Hawaiian Islands.

Lili'uokalani and her government had been deposed by a revolution in 1893. Five years later, the United States annexed the islands, making them a territory. Now, in 1909, at age seventy-one, Lili'uokalani was returning home after a trip to Washington, D.C., where she had tried to convince members of Congress to return to her some of the land confiscated during the revolution. Congressional members said no, as they would on each of her subsequent trips.

The seven-day passage to the islands onboard the SS *Siberia* was uneventful: It was good weather all the way, which must have pleased Helen Jaggar, who was making her first sea voyage anywhere.

Early on the last morning at sea, the island of Oahu came into view. The first landmark was Diamond Head, an ancient volcanic crater, its profile already familiar to travelers because it was already frequently reproduced in travel magazines. Next came a stretch of white sandy beaches lined by swaying coconut palms. This was Waikiki, already a tourist destination, home to the first major hotel in the islands, the five-storied Moana Hotel, which still stands today. And beyond Waikiki were the city of Honolulu and its harbor.

On the day the Jaggars arrived, Honolulu harbor was teeming with moving seacraft, as usual. Ships from a dozen distant ports were waiting to dock. Most were there to load bags of sugar, the islands' main export. Others, such as the *Siberia*, had passengers anxious to disembark, though, in these cases, most of the ships had come from Asia and were bringing workers for the cane fields.

As soon as the *Siberia* docked, there was a surge of people from shore onto the ship. Some were selling flowers. Others had coconuts or stalks of sweet cane to sell. Most were just there to greet passengers. Women were dressed in summertime dresses and carried parasols to protect them from the tropical sun. Men wore white linens and straw hats and canvas shoes. And everyone seemed to be waving, anxious to attract the attention of a relative or a friend.

The Jaggars had no one waiting for them nor did they have plans to meet anyone while in the islands. Their destination was Kilauea volcano, which meant an additional two-day sail to the island of Hawaii and the town of Hilo. They stayed the night in Honolulu. The next afternoon, they returned to the harbor and boarded a small steamer, the *Claudine*, which had originally been used to ship cattle between the islands, but was now carrying passengers.

The *Claudine* had no private rooms, only open berths where about three-dozen people could sleep. There was but one lavatory, used by both passengers and crew. It had no electrical lights, only kerosene. No food was served. It also rode low in the water, which meant seawater often washed across the deck and splashed across the faces of those who slept in the lowest berths.

On the first day of the run from Honolulu to Hilo, when the sea was calm as usual, the *Claudine* made its way along the leeward side of the islands. First there was Oahu, then Molokai, the steamer stopping at small ports to exchange passengers or to take on or deliver equipment for local sugar mills. By sunrise of the second day, the *Claudine* reached the old whaling port of Lahaina on the west side of Maui. Now it was a fast sail to the east side of the island—with a few stops—the captain anxious to reach the last port on the east side of Maui before noon so that he would have a full six hours of daylight to make it across the strait between Maui and Hawaii.

In the strait, the sea swells were always high and the wind was always strong. As it was often said by those who made the passage frequently, a ship never *sailed* between Maui and Hawaii, it *lurched* between the islands. One wonders how Helen Jaggar fared, this being her first time on a small ship in a rough sea.

If the captain was lucky, by sunset his ship was in the calm water of the leeward side of Hawaii. The *Claudine* made one stop at Kawaihae, then it was off again, this time around the north point of the island—and a twelve-hour sail along a windward coast.

Here sea swells were higher than those in the strait between Maui and Hawaii. Even under the best weather and sea conditions, the *Claudine* pitched and rolled wildly. Now almost all passengers were sick. But for those who could muster the strength in the early morning hours and stand just before sunrise, the view was spectacular.

The *Claudine* now passed close to high sea cliffs. Scoured into those cliffs in hundreds of places were narrow gulches where streams cascaded down, dropping the last few hundred feet into the sea. And above those cliffs, as one could see as dawn approached, was a wide swath of brilliant green, a continuous field of sugar cane.

And above the cane was a belt of dark forest that, in turn, as one looked higher, graded into barren upland.

That the trip was nearly over—one can almost hear the sigh of relieved passengers—was signaled by an abrupt end of the cliffs and the *Claudine*'s entrance into Hilo Bay. To the left, forming the distant horizon, was the broad arched profile of the volcano Mauna Loa, still active; the most recent eruption was just two years before the Jaggars visited. To the right, equally majestic, though more conical in form, was Mauna Kea, often with snow on its summit. Its latest eruption was in prehistory. And where the lavas of Mauna Loa and Mauna Kea met at the shoreline was the town of Hilo.

As any resident or visitor could attest, Hilo was a place more readily sensed than seen. Its scores of whitewashed houses and small cluster of commercial buildings were all but hidden by trees and thick foliage. In a few places church spires rose above the foliage. The population of Hilo in 1909 was about six thousand, making it the second largest community in the islands. The bay in front of the town was three times larger than the harbor that serviced Honolulu, which meant Hilo had the potential for growth. And the fact that Hilo was growing could be seen by Thomas Jaggar as he stood on the deck of the *Claudine* and sailed into the bay.

A stone breakwater was under construction. When finished, it would run for two miles, providing protection again ocean surges, allowing large ships to enter in almost any sea conditions. On shore was a new wharf, also under construction, and a new warehouse where thousands of cases of pineapples and tens of thousands of hundred-pound bags of sugar could already be stored, waiting to be loaded on the ships that were then in the harbor. There were stacks of lumber cut from nearby forests, ordered by the Santa Fe Railway to be used as railroad ties. And cowhides from island ranches ready to be shipped for tanning. And hundreds of barrels of tallow that would be used to make candles. Stacks of wooden crates held a variety of tropical fruit—bananas, mangoes, papayas—and were destined for distant ports. It was all a sign of economic prosperity, a prosperity that, though Jaggar could not have known it, would be a factor in bringing him back to the islands.

The *Claudine* landed its passengers—there were twenty-five that day—on a small wharf. Nearby was a locomotive, already at full steam. And behind the locomotive were two passenger cars, spartan affairs with straight-back wooden benches and open windows with only curtains to protect those inside from strong sun or from rain.

Though exhausted, passengers from the *Claudine* boarded the train. It was one more day of travel to reach the summit of Kilauea.

The tracks ran straight through fields of cane, the different heights and different hues indicating different stages of maturity. At five miles, the train plunged into a forest of moss-covered trees and hanging vines. "A burst of tropical jungle," wrote one traveler who had made the trip many years earlier, "I could not have imagined anything so perfectly beautiful." Here nature "rioted in the production of wonderful forms, as if the moist, hot-house air encouraged her in lavish excesses."

At ten miles from Hilo, the train left the forest and entered yet another cane field, making its first stop soon after at the Olaa Sugar Mill, the largest sugar mill on the islands which processed cane from the largest sugar plantation on the islands. Here some passengers left and others boarded. Then the train was off again, this time slower as it started the long climb up the side of the volcano. The rise was gradual, a steady increase in elevation indicated only by a growing coolness of the air. Taking advantage of the train's slower speed, the conductor often reached out with a gloved hand and snapped off a stalk of ripe cane. He then cleaned it with a machete and handed small pieces to passengers.

At twenty miles, three hours from Hilo, the train again entered a forest. This one was of huge *ohi'a* trees, a native hardwood, and giant ferns, some of the latter more than twenty feet high. Soon the train reached its last stop. Here volcano-bound passengers boarded a stagecoach, which Helen Jaggar remembered many years later as "a California buckboard affair, but higher and holding a dozen or more passengers." Here she picks up the narrative.

"The driver's name was Manuel, a Portuguese man, as lean as his team, swarthy and perched on the extreme front edge of his seat with his elbows out as though to help the poor beasts over the road. Instead of calling to them or swearing at them, he controlled them with a shrill whistle. When he wanted speed, the whistle would be the shrillest, most strident thing ever heard. When he wanted them to slack, his whistle became a caress."

The last miles were slightly steeper than those after the sugar mill and passed through a forest of scrubby ohi'a trees and thick brush. Open spaces covered with grass were now common. The elevation was nearly 4,000 feet and the air was distinctly cool. Occasionally, a crack could be seen on the side of the road where wisps of steam rose.

As the road finally leveled out, there was a gradual turn to the right, then one to the left. After the last turn was made, a large two-storied clapboard building came into view. The sides were painted a vivid yellow; the roof was bright red tin. A row of bushes, always in bloom, lined the road that led up to a portico where a short robust man with a wide handlebar mustache stood. He was Demosthenes Lycurgus, proprietor of the hotel. And this was the famous Volcano House at Kilauea volcano.

Lycurgus greeted the new arrivals as each person stepped down off the stagecoach. Then, with everyone assembled and the bags unloaded, he led them up a short flight of stairs to a verandah. From there, they passed through double glass doors, the panes etched with floral designs, and continued into the hotel lobby. A long elegantly carved reception desk ran along one wall, made of a local mahogany known as koa wood. Through one of the doors, as Lycurgus directed the attention of his new guests, was a ladies' parlor with a piano. Through another was the smoking room for men. And through a third, off to the left of the lobby, was the main parlor.

Here the floor was covered with fine mats and rugs. Furnishings were mostly sofas and chairs made of rattan. The most striking feature was the far curved wall of windows.

Through those windows one could see that the hotel was built near the edge of a cliff. And that extending away from the base of the cliff was a great sunken plain, devoid of vegetation. Then, if one continued to stare and study the scene, one could see out on the plain, two miles from the hotel, the sharp outline of a circular crater. That, explained Lycurgus, who took pleasure in introducing the feature to new arrivals, was Halema'uma'u, the home of the volcano goddess Pele and the location of the famous lava lake.

———•———

The first written use of the name Halema'uma'u was by Count Pawel Edmund Strzelecki, who introduced himself as a Polish nobleman, though no legal claim to the title was ever established. He arrived in the Hawaiian Islands in 1838, visited Kilauea late that summer, wrote a letter about his visit, which was published in the newspaper *The Hawaiian Spectator*, and then left the islands. It was in the letter that he wrote the name of the crater as *ale mau mau*, which he said meant "the great Whole of the volcano."

Hale-ma'uma'u is the accepted name of the crater today, but its meaning is very different from the one given by Strzelecki. *Hale* is the Hawaiian word for "house." *Ma'uma'u* refers to the *'ama'u* fern, a small fern that grows in abundance on Kilauea and is one of the first plants to appear on a new lava flow. And so, the name *Halema'uma'u* literally means "fern house." But there is another, more subtle meaning that is difficult to translate. It refers to a continuing battle between the volcano goddess Pele and the pig god Kamapua'a. In the fight, Pele uses the fire of the volcano as a weapon. As a creature of the forest, Kamapua'a relies on ferns to entangle and trap her.[*]

On a different note and to clarify a common misunderstanding, the great sunken plain at the summit of Kilauea is not a crater. It is a *caldera*, a term introduced to geology by Clarence Dutton after a

---

[*] To complicate the meaning of *Halema'uma'u* further, it was also the name of one of six priests who came in ancient times to destroy Pele and extinguish her volcanic fire.

visit to Kilauea in 1882. He chose the Spanish word *caldera*, which means "cooking pot," to signify that the plain at the summit of Kilauea formed by collapse, much like oatmeal does when it is cooked in a pot and collapses when removed from heat. In this case, as Dutton envisioned it, the sunken plain is the top of a large crustal block that settled downward when a buried reservoir of molten lava was tapped and the lava was drained away and erupted from the volcano.

A mention must also be made of the type of volcano that Kilauea is; why its appearance is so different from the common notion of a volcano as a solitary cone with steep sides, such as Vesuvius or Mount Pelée. In geologic language, those are *stratovolcanoes*, a term first used in 1866, and owe their shapes to the fall of exploded material and to the eruption of highly viscous lava that can flow only short distances before cooling and turning solid. In comparison, the lava that was, and still is, erupted at Kilauea—and Mauna Loa—is highly fluid and forms broad sheets that can flow for many miles. Evidence of these sheets is clearly seen in the caldera wall at Kilauea that consists of layers of thin lava flows.[*]

Because of their broad profiles, Hawaiian volcanoes were originally known as *dome volcanoes*, also first used in 1866, though a more picturesque name has since been adopted. In 1911, two years after Jaggar made his first trip to Kilauea, a German geology student, Hans Reck, went to Iceland to search for a lost friend. He never found the friend, who apparently drowned when his boat overturned in a lake, but, from his wanderings in Iceland, Reck did contribute a new term to geology, *Schildvulkane*, or "shield volcano," suggesting the low profile of Icelandic volcanoes resembled a

---

[*] The reason for the higher viscosity of lava erupted at Vesuvius and Mount Pelée—and other stratovolcanoes—is the higher silicate content of the lava, an idea proposed in the 1880s, but not set on a firm foundation until the 1930s when the new technique of X-ray crystallography showed that silicon and oxygen atoms were arranged in long chains that can entangle, creating a lava of high viscosity. In comparison, lava erupted at Kilauea have relatively low amounts of silicon and atoms and do not form the long chains.

warrior's shield laid horizontally. Reck even took the next step and applied the new term to Hawaiian volcanoes, noting, as many had before him, that Icelandic and Hawaiian volcanoes were "analogous in all essential characters," a view that that geologists continue to echo today.

———•———

The need to return soon to Honolulu to catch the next scheduled passenger ship to Japan limited the Jaggars' stay at the Volcano House to three days. On each of those days, Thomas Jaggar made the long trek on foot to stand at edge of Halema'uma'u and see the lava lake. On the third and last day, he was accompanied by his wife.

The path to the crater began a hundred yards from the front of the hotel. Here one made a descent down the cliff, the path following a zigzag course along a series of steep slopes interrupted by an occasional broad shelf. It was a good foot trail maintained by hotel workers. At the bottom, 400 feet below the level of the hotel, one stood at the edge of a vast plain of congealed lava.

Now it was a straight course over two miles of hard lava. In places, the surface undulated, the black lava having frozen into giant waves. Elsewhere, the lava looked as if it had swirled as great eddies. Cracks where steam was rising had to be jumped or bypassed. The surface itself was one of black glass—"not deep black," Helen would write, "but an iridescence of darkest blues and purples and greens, which might have beauty if it weren't so devilish." In spite of the frequent foot traffic, the trail remained rough and uneven beyond belief. And the volcanic glass was still sharp, capable of wearing down all types of footwear, given enough passages back and forth along the trail, even heavy hobnailed boots.

About three-quarters of the way to the crater, a little off a straight path, were the famous "hot postcard cracks" where visitors stopped to singe postcards that were later mailed to friends. The technique to singe a card was a simple one.

A person got upwind of a chosen hot crack, so as not to inhale unpleasant fumes, then, with a wooden stick that had been split at one end, inserted the card firmly and thrust the card into the crack to a depth of four or five inches. Within seconds, the card would either be toasted brown or burst into flames. If the latter happened, a new card was inserted and the whole procedure tried again. For those who had trouble mastering the technique, there was never a shortage of cards or wooden sticks. Each morning, hotel proprietor Lycurgus sent a boy with several sticks and hundreds of cards, instructing him to sell the cards a dozen at a time for 50 cents.

Just beyond the postcard cracks, the trail became steeper and the surrounding rocks were more jagged. Closer still and the surface was a jumble of rocky plates, similar in arrangement to the giant plates of ice that form where ice jams form on a frozen river. But here the rocks were hot to the touch, making it difficult to stand still in one place. Now progress was slow, the edges of upturned plates sharp enough to lacerate skin.

Within a few hundred feet of the crater rim, golden strands could be seen floating in the air. These were strands of volcanic glass, known as "Pele's hair," formed when wind blew through and caught molten lava as it spurted into the air. Closer still, and tiny black beads, teardrop in shape, filled the air. These, too, were of volcanic glass—tears of the goddess.

At last, the edge of Halema'uma'u was reached. In April 1909, when the Jaggars visited, the crater was nearly circular in form, about 1,500 feet in diameter. A hundred feet down was a wide ledge of black solidified. Another hundred feet down, covering the bottom of an oval inner pit, was a pool of molten lava, seething and roiling.

"The pit of boiling lava within the crater of Kilauea is a spot of infinite excitement and fascination," Helen Jaggar would remember of the experience. "Excitement because of the seemingly imminent danger of having the rim on which you are standing suddenly crumble and catapult you into the pit below, if you don't lose your balance and fall in anyway—and fascination because of the complete instability of the lava below."

"There isn't a second's pause in the ebb and flow of the lava," Helen noted, "the satiny, black-grey crust forms only to be broken at once by the lightning-like cracks in the surface exposing the most livid-flame colored molten lava." She continued. "All around the edge of the lava lake rose sheets of steam, tinted, by the color of the lava, to the most lovely shades of pink from delicate shell pink to deep rose."

In places, red incandescent lava rose up and broke through the black-grey crust, causing the crust to rip apart and drift away as great detached sheets. In one corner was a jet of lava that spouted at nearly regular intervals of about one minute. With each spout, it flung molten lava nearly a hundred feet into the air. Demosthenes Lycurgus named it "Old Faithful" in recognition of the famous geyser at Yellowstone—and to attract visitors to the volcano and to his hotel.

Helen Jaggar discovered that she had to frequently retreat and sit well back from the edge to keep the radiant heat from the lava lake from burning and blistering her face, crawling forward only occasionally to peek. She and her husband watched the lava lake for more than two hours. Hardly a word was spoken between them. They sat there, she recalled, "with absolute no realization of the passage of time."

Her husband also recorded the scene. "Looking down into the crater of this active volcano," he wrote, "the pit presents an orange, almost white-hot bed of undulating lava, pounding on the rock shores and rolling like surf." He also noted the lake was covered with a thick black crust through which a few fires "flowed with an unearthly gleam." The crust consisted of a half-dozen sheets, each one slowly shifting. At times, one of the sheets would pitch up suddenly like a giant ship, then slide under the surface of the lake. That would set the entire surface of the lake in motion, transforming what had been a relatively placid dark lake into a mass of red fire as more sheets tipped over and sank, causing the surface to boil at dozens of points, each one spraying molten lava and filling the air with Pele's tears and hair.

He also had measurements to make.

Thomas Jaggar hustled off, leaving Helen alone. He had brought a Bristol pyrometer, a device that could measure the hot temperature of gases in furnaces. He used it to measure the temperature of red-hot cracks around the edge of Halema'uma'u, finding that the hottest ones registered nearly 300°C (570°F)—the same as the hottest temperature he had measured in piles of volcanic ash at Mount Pelée.

After they returned to the hotel, he recorded his temperature measurements in the *Volcano House Register*, a book kept at the hotel for guests to record their descriptions and their impressions of the volcano. In the entry for April 7, 1909, he acknowledged that, in making the temperature measurements, he "was assisted by Mrs. Jaggar."

———•———

The Jaggars left the summit of Kilauea and the Volcano House the next day and retraced their route back to Hilo, then to Honolulu. In a few days they would board a ship and sail for Japan. In the interim, Thomas Jaggar was asked to give a lecture about his work on volcanoes.

The lecture was given on Wednesday, April 14, at the University Club of Honolulu. Club members were men who had attended at least one year of college and who now held a position of prominence in the islands. Some were politicians; most were businessmen. Included in the audience the afternoon of his lecture was the Territorial Governor, Walter Frear, and the Chief Justice of the Territorial Supreme Court, Alfred Harwell. Sidney Ballous, an Associate Justice of the Supreme Court of Hawaii, introduced Jaggar. He and Jaggar had been classmates at Harvard.

Jaggar had given the lecture many times, adding information about other geologic calamities when they occurred. He began this particular lecture by showing lantern slides of himself at Martinique and of eruptive clouds rising from Mount Pelée. He told of the work being done by Perret at Vesuvius and of his own adventures in Alaska. He showed slides recently sent to him by Perret of the destruction of the city of Messina in southern Italy. An earthquake had shaken the city on December 28, 1908, killing more than a hundred thousand people—more than half the city's population. At the end of the lecture, he turned to the audience and told them: "I come here tonight to make a missionary appeal."

"What was the cause of the Messina earthquake or of the recent one in San Francisco?" he asked. He said that he did not know. Nor did anyone. The ignorance came from the fact that scientists, such as himself, rushed to a scene only *after* a geologic disaster has occurred. What was needed were geonomical observatories that could record what was happening inside the earth. "That doesn't mean observatories on the tops of volcanoes with heroes to sit on the edge of craters all day," he said. Instead, it required people who would devote their lives to the careful and continuous recording of activity, using the best scientific equipment then available. Each observatory would keep a specific region of the world under surveillance. And the people who worked there would issue a warning if a geologic disaster—an earthquake, an eruption or, perhaps, the approach of a giant sea wave—seemed imminent. Such work would be historic and represent a new type of profession, one that Jaggar described as being "a missionary of science."

The first such geonomical observatory had already been established at Vesuvius. The second one, Jaggar announced, would be near Boston, operated by MIT and funded by a recent gift from the Whitney family of Boston.

After the lecture, a man with a balding head and cherub face introduced himself to Jaggar. He was Lorrin Thurston, the owner and publisher of the *Pacific Commercial Advertiser*, the largest newspaper in the islands. He asked whether Kilauea might be a better place to establish the next observatory. Jaggar said that it would. "Is it a matter of money?" Thurston asked. Jaggar answered that it was, largely, but it also required persuading the Whitney family and MIT administrators to shift their attention to the Hawaiian Islands.

Thurston suggested that Jaggar stop in Honolulu on his return from Japan so that they could talk more about what might be done at Kilauea.

The Jaggars stayed in Japan two months. Helen spent her time in Tokyo and Yokoyama. Her husband went off, having time to travel to and climb only two of the countries more than a hundred volcanoes.

The first one he visited was Asama, sixty miles northeast of Tokyo, one of the most active volcanoes in Japan. He climbed the steep slope, led by Japanese hosts. When they reached the summit, the crater was completely clear of smoke and steam, something, his hosts said, that had not happened in years. Jaggar took several photographs, one showing a large swirl of solidified lava at the crater bottom. The most recent explosion of Asama was a minor one the previous September. The next one would occur the next December.

The second volcano was Tarumai far north of Tokyo on the island of Hokkaido. Tarumai had been quiet since 1896. On March 30, 1909, six weeks before Jaggar's visit, a tremendous explosion took place, throwing out ash and lava bombs. Two weeks later, another explosion occurred, this one described by Japanese observers as "in full violence." Those who climbed the low cone four days later saw that new lava, in the shape of a giant dome, a quarter of a mile wide and more than six hundred feet in height, had risen from the crater floor.

Jaggar was at Tarumai three weeks after the second explosion. The dome reminded him of the craggy rocks he had seen on Bogoslof. On the day he visited, the volcano was quiet, only steam rising passively from the base of the dome. Occasionally, rocks could be heard falling off the dome. Jaggar decided to descend into the crater.

He approached the dome, carrying his Bristol pyrometer. He found a crack in the side of the dome and thrust the end of the pyrometer into it. The temperature was 855°F (457°C), the hottest he had measured at any volcano. As he would write, there was enough heat a foot from his face to ignite wood.

The visits to Asama and Tarumai added to his experiences at volcanoes. But the lasting benefit of the trip to Japan came after he returned to Tokyo and met Fusakichi Omori.

Omori was a shy, diminutive man of extraordinary intellect. Unlike most Japanese of his generation, he had traveled widely. He had been to Europe and the United States. He spoke English, French and German fluently. And he was remarkably prolific. In 1908 he published a catalogue of earthquakes that had occurred in Japan since the 5th century, listing 9,628 events, noting that 621 had caused some damage, indicating that Japan was struck, on average, by a destructive earthquake every two or three years. He had formulated what is known today as Omori's law, which states how the rate of aftershocks decrease after a major earthquake. He had also invented what would soon become the first widely used instrument to record earthquakes.

To understand how the Bosch-Omori seismograph worked—Bosch referred to the firm of J. & A. Bosch of Strasbourg, Germany, where the instrument was commercially produced—think of a door that is free to swing on its hinges. When the ground vibrates, even slightly, the floor and walls of a building also shake, including the hinges of a door, but the door itself, because of its inertia and because it is hinged, tries to stay in the same position. If a stylus is attached to the bottom of the door at its unhinged end and a sheet of paper is placed under the stylus, then, when the ground shakes, an autograph of the shaking is recorded on the paper. The Bosch-Omori seismograph works in the same way.

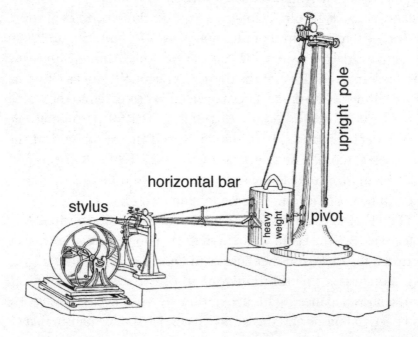

One end of a rigid horizontal bar—the bar represents the door—is attached to a sturdy upright pole—the hinges—in such a way that the bar is free to pivot. A stylus is placed at the other end. A sheet of paper is wrapped around a cylinder that rotates slowly and the cylinder is placed under the stylus. What happens when the ground vibrates during an earthquake? Just before the earthquake, everything is stationary and the stylus traces a straight line of the rotating cylinder. Then, as the ground starts to vibrate, so does the upright pole and the cylinder, but the horizontal bar and the stylus tend to remain stationary because of the inertia of the bar, which is augmented by a heavy weight, and because one end of the bar is free to pivot again the pole. And so, as the cylinder vibrates under the stationary stylus, the line traced out on the paper is no longer straight but wiggles back and forth giving a permanent record of ground shaking during an earthquake.

It was an ingenious device, purely mechanical, and so no electrical power was required, which meant it could be used in remote locations. By 1909, Bosch-Omori seismographs were in use at more than a dozen places around the world, including six in the United States. And, as Omori was fond of pointing out, his seismograph could do more than just record earthquakes. It could also record a slight tilting of the ground.

Think again about a hinged door that is free to swing. If a building settles unevenly, so that the floors and walls tilt, that will cause the hinged door to swing and hang at a new position. In the same way, if the ground tilts slightly, then the upright pole of the Bosch-Omori seismograph is no longer at vertical and that will cause the rigid horizontal bar to swing to a new position, shifting the position of the stylus on the paper.

It was, he would write of his trip to the Pacific, as if "everything within me converged." There were businessmen in the Hawaiian Islands who seemed ready to fund his scientific work. There was a volcano on the island of Hawaii that was erupting continuously and could be approached closely. And here, in Japan, there was a new scientific instrument—the best yet devised to record

earthquakes and to measure the slight, more permanent movement of the earth's crust—that might reveal unseen activity
happening deep inside a volcano—and thereby forewarn of eruptions. Jaggar was now ready to hustle back—not to Boston—but
to Kilauea and start the work.

Helen and her husband returned to Honolulu on June 7. Helen
left the next day for San Francisco, where she continued on back to
Boston. Her husband remained in Honolulu and called on Lorrin
Thurston.

———•———

Jaggar was received with much enthusiasm when the two men met
again. Thurston took him immediately to a private meeting with
Governor Walter Frear, Thurston's former law partner, who said he
"wholly supported" a science station at Kilauea. Then Jaggar, Thurston and Frear went to see John Gilmore, the president of Hawaii
College, who also thought a science station at Kilauea was a good
idea and promised his support. He also offered Jaggar a professorship if he decided to move to the islands and run the station.
Then Jaggar and Thurston went to Bishop Museum and called on
the director, William Brigham, who also gave them a promise of
support, Brigham suggesting that any lava samples or earthquake
records that Jaggar might collect at Kilauea be sent to the museum
for permanent storage.

A few days later, Jaggar spoke at an afternoon meeting of the
Honolulu Chamber of Commerce, telling its members of the plans
for Kilauea. At the end of the lecture, Thurston rose and addressed
the audience. He said that, in addition to the scientific benefit, there
was also "the purely commercial advantages of securing for Kilauea
a volcano observatory." Moreover, the inclusion of an institution
such as MIT would advertise the islands in a way that "was impossible by any other means."

A prepared statement was then handed out to members of
the Honolulu Chamber of Commerce. In it, Jaggar, as MIT's

representative, "tentatively offers" to carry out all work necessary to establish a volcano observatory and to pay for it with a financial gift recently given to MIT by the Whitney estate of Boston. For his part, Thurston pledged to collect $5,000 in annual subscriptions from those present to operate the observatory. By the end of the meeting, Thurston had collected nearly half the amount.

Jaggar spent another month in the islands, shuttling twice between Honolulu and Kilauea and the lava lake. His ducks were now in a row: He had firm local support. By August he was back in Boston ready to put the final piece in place.

He met with MIT President Richard Maclaurin and laid out the advantages of working at Kilauea. An active volcano would be a better place than Boston to use the $10,000 gift from the Whitney estate to establish a seismological station. Maclaurin promised to take up the matter with MIT's Executive Committee.

The committee met in October, Maclaurin presiding. After careful deliberation, so a later report said, Maclaurin and the members of the committee came to a decision. They saw no provision in the Whitney gift to spend the money anywhere except near Boston. Furthermore, as to the specific question of anyone at MIT doing work in the Hawaiian Islands, Maclaurin wrote directly to Jaggar, saying "the Committee did not think it opportune at this time to embark on such a project."

And so the question seemed settled. There would be no volcano observatory at Kilauea, at least, not one that involved MIT. It was a decision that probably pleased Helen Jaggar—who would have been even happier if she was not so furious with her husband.

———•———

On their trip to Japan, Helen had left Boston two months before her husband did, Helen spending the time in San Francisco with her family. Meanwhile, as he had promised, her husband had searched for a house to buy. But, before he started the search, he did something else. He bought a new car.

This was actually his second car. The first had been a steam-powered Orient Roundabout, manufactured by the Waltham Company in nearby Waltham, Massachusetts. Jaggar purchased it in 1905, making him one of the first in Massachusetts to own an automobile. In essence, it was a motorized wooden buckboard with a single bench seat for two people. It had a one-cylinder, four-horsepower motor. A tiller was used to steer the machine. By 1909, it was a greatly underpowered vehicle at time when automobile technology was developing rapidly.

The automobile purchased by Jaggar in 1909 was a Peerless Model 19 Touring Car, probably the most luxurious vehicle of the day. It had front and back seats. Steering was done by turning a wheel. The body was all steel except for a retractable leather top. It came with a windshield and fenders. Inside were two dials: a speed-ometer and a clock. And it had a powerful motor: a four-cylinder, 34-horsepower engine, water-cooled and fueled by gasoline.

It was in this new automobile that Jaggar drove the countryside around Boston, searching for a house. When he found one to his liking, he stopped and inquired if the owner was interested in selling. He finally found an owner who was. The house was a stone mansion that sat atop a hill in south Brookline with magnificent views of Boston and Cambridge. Around it were pastures and apple orchards. Adding to the attractiveness, the stone near the front door had the year "1792" carved into it, suggesting that some of the building materials, if not the entire house, dated from the Revolutionary War.

There were fifteen rooms, several with fireplaces. Jaggar had their furnishings moved, though only had enough furniture to fill three of the rooms. And so, he added a grand piano in one of the vacant rooms. Three people would be hired to run the household. There was 22-year-old Nellie Leonard of West Virginia who would serve as the maid and the much older Lucy Johnson of Virginia who would be the cook. To work as handyman and gardener and to serve as Mrs. Jaggar's chauffeur and drive her in the Peerless wherever she might want to go was Joseph Bell of Georgia. This would be the

Jaggar country estate—and Helen disliked it as soon as she saw it, regarding it, for the remainder of her life, "as just another of Mr. Jaggar's crazy ideas."

Theirs was the only house on the crest of a steep hill. The nearest neighbors were an old man and his housekeeper who lived in a shabby building at the base of the hill. The next nearest neighbors were a family who, for a reason she never disclosed, Helen decided she did not care to know. She wanted to live in Boston. She wanted to be surrounded by society.

Not only did her husband resume his travels soon after he returned from Japan, leaving her to live in the stone mansion, but he had bought a second, smaller house. It was a summertime beach house in West Yarmouth near Lewis Bay, seventy miles south of Boston. In her eyes, the community "was a queer little town not much more than a store and not very attractive but one of the 'cheap' places the Jaggars are used to."

As Helen would write: "We returned in the fall of 1909 and had a most difficult time of it."

It was soon to get worse.

CHAPTER EIGHT

INTO THE CAULDRON

Lorrin Thurston was the grandson of Asa and Lucy Thurston, members of the first group of American missionaries to arrive in the Hawaiian Islands. As a child, young Thurston learned to speak Hawaiian fluently. As an adult, he signed letters and had close friends call him *Kakina*, the Hawaiian name given to his grandfather.

In 1880, at age nineteen, he entered Columbia Law School in New York where one of his classmates was Theodore Roosevelt. Both men attended the school for one year. Neither man graduated. Both came away with a determination to fight government corruption.

The Hawaiian Islands was then an independent country, a kingdom, and the ruling monarch was King David Kalakaua. In 1875 Kalakaua signed a treaty with the United States that allowed certain Hawaiian goods, mainly sugar and rice, to enter the United

States tax-free. In exchange, the Kingdom of Hawaii granted the United States access to Pearl Harbor as a military naval base. The treaty brought economic prosperity to the islands, which strengthened Kalakaua's political control, allowing him to form and dismiss cabinets and to spend money at will. It was the growing autocracy—and a desire for the islands to have a closer tie with the United States—that caused Thurston to enter politics.

He became a political firebrand, writing editorials and making public speeches accusing Kalakaua of corruption. In 1886 he sought, and won, election to the island legislature. He then joined a secret organization of like-minded individuals who, on June 30, 1887, acted, forcing Kalakaua to accept a new constitution, the so-called Bayonet Constitution, that stripped the king of most of his executive powers and put the powers in the hands of cabinet members. They then forced Kalakaua to make new cabinet appointments, which included naming Thurston as the Minister of Interior.

The brash 29-year-old now had considerable control over the daily running of the kingdom. He was in charge of the immigration office. He decided on the location of new harbors and where new bridges and government buildings would be constructed and which new roads would be built.

In 1889 he began what was, up to that time, the largest construction project in the islands: a thirty-mile-long carriage road that would link Hilo and the summit of Kilauea. A few months after construction was begun, he made a trip to inspect the new road. It was on that trip that Thurston decided to buy the Volcano House hotel.

The first commercial enterprise to cater to visitors—and to call itself the "Volcano House"—opened in 1846. It consisted of a small thatched hut located close to the caldera rim. A second, larger Volcano House, also a thatched hut, was built in 1866. It had a single main room with a brick fireplace. High grass partitions separated bedrooms from the main room. This was the Volcano House where Mark Twain stayed for a few days and wrote the short story "A Strange Dream" about Kilauea and the search for ancient Hawaiian bones. A third Volcano House, the one that Thurston would buy, was completed in 1876. This one was constructed of wood planks cut from the surrounding ohi'a forest. It, too, had a brick fireplace, as well as wooden doors with brass knobs and window frames with glass panes shipped from San Francisco. There were six guest rooms. Each one contained a single bed. Up to three guests were assigned to a room.*

As soon as he purchased the hotel, Thurston decided to expand it. He added a dozen rooms, a new kitchen and a dining room and separate parlors for men and for women. Kilauea had been erupting vigorously for years when Thurston bought the Volcano House—a lava lake roiling and surging within Halema'uma'u—and so he had no trouble attracting investors. On May 5, 1891, he was in Honolulu and announced the incorporation of the Kilauea Volcano House Company. That night the floor of Halema'uma'u collapsed and the lava lake drained away. Three weeks later, much to his relief and the relief of his investors, molten rock reappeared—and visitors kept coming to the volcano.

Earlier the same year King Kalakaua died and his sister, Lili'uokalani, became the ruling monarch. The events that followed have been told many times. Here just a few key ones will be summarized.

Soon after Lili'uokalani became queen, the islands' economy collapsed. The economy was based on Hawaiian sugar entering the United States without payment of an import duty. It was part of a reciprocity treaty that allowed the United States exclusive use of Pearl Harbor for its Navy ships. But Congress eliminated the favored trade status of the Kingdom of Hawaii, which collapsed the islands' economy.

Thurston was not yet involved in the sugar industry, but he was involved in promoting a new industry—tourism. In May 1891, he offered the first package tours to the Hawaiian Islands: a weeklong stay at his Volcano House. The tour was so successful that, in August

---

* This building is still at Kilauea and houses a gallery for the Volcano Art Center.

1892, he and others organized the Hawaii Bureau of Information, a forerunner of today's Hawaii Tourist Bureau. A month later he was in the United States promoting Hawaiian tourism. He took the opportunity to visit Washington, D.C., and meet with several government officials, asking what their reaction would be if action was taken against the Hawaiian government and Queen Lili'uokalani was ousted. According to Thurston, those who he met said they would be "exceedingly sympathetic" if such action was taken.

On January 14, 1893, after closing a contentious legislative session—the Queen's supporters and her detractors often trading accusations of favoritism and corruption—Lili'uokalani announced a new constitution that restated her right to rule. Thurston and others saw their moment had come.

Three days later, on January 17, they confronted Lili'uokalani who, under protest, transferred control of the government to them. Thurston then raced to Washington, D.C., to ask for annexation. But the politics of the United States had changed. The new president, Grover Cleveland, elected the previous November, was not sympathetic to a Hawaiian revolution and thought the revolutionaries should give control of the islands back to Lili'uokalani. Disregarding his advice, the Republic of Hawaii was proclaimed on July 4, 1894. On that day, Thurston and his family were en route to Hilo and Kilauea. He planned to settle at the Volcano House for a long stay.

The volcano was in a spectacular eruption when he arrived. The level of the lava lake was the highest ever recorded, twelve feet *above* the rim of Halema'uma'u, held high by a natural levee of solidified lava. People were rushing from Hilo to see the activity. The Volcano House was filled with guests. By July 10, the lake surface was thirty feet above the rim. Then, the next morning, when Thurston awoke and looked out the hotel's windows, all he could see were great clouds of dust rising from the crater.

He and dozens of guests rushed to the crater. He would stay the remainder of the day and most of the following night. At the end, after the dust cleared the next morning, he saw that the crater floor was more than six hundred feet down from its previous level, and that no molten rock was in sight. The entire lava lake had drained away, the molten lava flowing back into the volcano through a crack deep inside the crater. In fact, molten rock would seldom be seen at Kilauea for the next several years. And so the Kilauea Volcano House Company became a dead business.

In 1898, the politics in Washington, D.C., changed again. The United States was at war with Spain and needed a way station for its Navy ships in the Pacific. And so, on July of that year, Congress annexed the Hawaiian Islands. Four months later, now assured of a stable market for Hawaiian sugar, Thurston organized the Olaa Sugar Company near Hilo. His would be the largest sugar plantation and the largest sugar mill in the islands.

From the beginning, the Olaa Sugar Company lost money. Some of the loss was from a drop in sugar prices. Most of the loss was due to an extraordinary run of wet weather that caused the stalks to rot. Some people said the company was failing because of the name. "Olaa" signified a revered place: a forest that had been used since ancient times for the gathering of medicinal plants and for making the Hawaiian cloth, *tapa*, from the *mamaki* tree which grew in the forest in profusion. But, now, where the forest had stood, Thurston had his sugar plantation.

Company creditors called in loans. Thurston defaulted twice. The third time the creditors stood firm. And he was forced to sell another failing property: the Volcano House.

He sold the hotel to George Lycurgus, a Greek immigrant who had been a supporter of the Hawaiian monarchy. In fact, Lycurgus had been part of a counterrevolution, an act that led him to being jailed for a year.

After repairing the long neglected hotel and improving the furnishings—he replaced the old corn-husk mattresses with factory-made, straw-filled ones—Lycurgus reopened the Volcano House on Valentine's Day in 1905. A week later, Kilauea roared back to life.

"Two fissures are pounding out hot lava," reported one eyewitness to the unexpected eruption. Another wrote: "Lava overflowing, covering apparently about an acre. Apparently increasing."

That night, for the first time in more than ten years, the Volcano House was filled to capacity. There were so many guests that some slept on the new billiard table in the men's parlor. Others slept on the floor or in hallways, anywhere that could be found. By the third day of the eruption, guests were arriving from Honolulu.

Thurston's fortunes also changed after he sold the Volcano House. The price of sugar rose and the weather improved. His sugar plantation and mill was finally showing a profit. And he used the money to expand his influence. He successfully lobbied Congress to build a breakwater in Hilo Bay and to contract with one of his new companies to construct it. The stone came from a quarry owned by Thurston and was carried by a railroad company owned by Thurston. The Santa Fe Railway contracted with him for railroad ties, which were cut at a new lumber mill he owned. And he was building a new wharf and a new warehouse at Hilo Bay.

He also began to lobby Congress to make the summit of Kilauea volcano into a national park. That way the unique character of the volcanic landscape would be preserved for generations. A national park would also preclude any further commercial development at the summit. Only the Volcano House would exist. And he was already planning on how to get ownership of the hotel back from Lycurgus.

It was during this contentious political, economic and volcanic back-and-forth in Hawaii that Jaggar appeared with his idea to establish a volcano observatory at Kilauea. Such an observatory would add to the attraction of Kilauea as a tourist destination. By then, Thurston also owned the *Pacific Commercial Advertiser*, the largest newspaper in the islands. He immediately saw a kindred spirit in Jaggar, as well as another means to expand his influence, and would use the newspaper to promote Hawaiian tourism and his several businesses—and the scientific work Jaggar was proposing for Kilauea.

———•———

Back in Boston, Jaggar did not consider MIT's decision to keep the money from the Whitney estate for a project near Boston to be a final one. So he went to see the person who ran the estate directly, Anne Rebecca Whitney.

A tiny woman with short curly silver hair, Miss Whitney, born in Watertown, Massachusetts, in 1821, had accomplished many things. She had been an abolitionist and was now an ardent suffragist. She was a noted sculptress of considerable and controversial reputation. Her first major work was a marble statue that showed a clothed Lady Godiva ready to disrobe. Another early work was entitled "Africa" and showed a recumbent female figure rising and awakening from freedom. Whitney was an advocate of the public financing of education for former slaves. She also campaigned for the conservation of forest long before Theodore Roosevelt entered the White House and made the cause a popular one.

In 1904, after the death of her aunt, Caroline Rogers Whitney, Anne Whitney inherited the family fortune. She used it to support the performing and visual arts. She underwrote performances of orchestras and the production of stage plays. She traveled to Europe where she purchased paintings, donating many of them to American museums.

In 1906, after the San Francisco earthquake, she asked one of the trustees of her estate, Charles Stone, what she might do to help prevent such future disasters. Stone was then also a member of MIT's Executive Committee and he suggested a donation of $750 to purchase a Bosch-Omori seismograph for MIT. One was purchased and it arrived in November 1908. Then, after the Messina earthquake and the deaths of more than 100,000 people, she again asked what she might contribute. This time Stone suggested a donation of $10,000 to establish a laboratory at MIT for the study of earthquakes near Boston. Again, she donated the money. Now Jaggar went to see her and ask for more.

As expected, he was charming at their meeting, but Whitney was not one who succumbed to charm. She wanted to know particulars. And so Jaggar laid out his plan.

A laboratory to study earthquakes should be established near Boston. And it would be the headquarters for a worldwide network of stations located at the most dynamic parts of the planet. The next station should be at the summit of Kilauea where earthquakes were numerous and where volcanic activity was continuous. The next logical place was Alaska, then in Panama where the United States was building a canal, then the Philippines and Puerto Rico, recently acquired possessions of the United States where earthquakes were also numerous.

Whitney asked how much money was needed.

"$100,000," he answered.*

Her reaction is unrecorded. But, the next day, MIT President Maclaurin and MIT Executive Committee member Stone went to see her.

Stone later wrote to Jaggar saying that he and Maclaurin "discussed the whole matter with her." She agreed to increase her latest donation to $25,000 to establish "a Research Laboratory of Physical Geology at MIT." As to work at Kilauea, Stone wrote that he still had "some doubt about the wisdom of an Hawaiian observatory."

And so the matter rested—until nature intervened.

On April 13, 1910, an earthquake shook central Costa Rica, destroying many buildings and killing nearly a thousand people in the capital city of San José. Three weeks later, on May 4, a stronger earthquake struck the same region. It came in the evening when people were at their dinner tables. To at least one who survived, the second shaking had come on "like the snap of a whip." Again, nearly a thousand people died. The United Fruit Company, which owned much of the national debt of Costa Rica, was headquartered in Boston. Its president was Francis Hart. Hart was also the Treasurer of MIT.

After the second earthquake, Hart summoned Jaggar to his office and asked him to go to Costa Rica and determine if another disaster were imminent. The United Fruit Company would pay all of his expenses.

Jaggar traveled by train to New Orleans where he boarded one of the snow-white steamers owned by the fruit company. Six days

---

* About $2,500,000 in today's dollars.

later, he arrived in Limón, Costa Rica. From there, it was another train trip to San José where he spent the next week making a map that showed the degree of earthquake damage. That was followed by a week climbing two nearby volcanoes, Irazú and Poas.

After Costa Rica, he went north to Guatemala where he tried to find passage to the Hawaiian Islands, but the closest steamer that sailed to the islands was in California. And so he abandoned the attempt and headed south to Panama where a canal was under construction. After Panama, he returned to Boston.

He had been away for only a month, but, during the month, the attitude at MIT toward him working at Kilauea had changed. Maclaurin and Stone were no longer opposed to such work, but they wanted it limited. The $25,000 from the Whitney estate would be an endowment. And Jaggar could draw on the first year of interest on the endowment—about $1,000—to work at Kilauea.

It was a small sum. But it was a start. And Jaggar set upon deciding how to spend it.

———•———

If one consulted one of the popular geology books of the time—Sir Archibald Geikie's *Text-book of Geology*, which some still consider to be one of the best geology books ever written, or Thomas Bonney's *The Story of Our Planet* or John Wesley Judd's *Volcanoes: What They Are and What They Teach*—one would find a consensus of opinion as to what was the most important measurement to be made on an erupting volcano: the temperature of molten lava.

Thermodynamics—the branch of physics concerned with heat and temperature and how they are related to energy and work—was still in its infancy, its laws having been in development since the 1750s. The originators of thermodynamics naturally gravitated toward volcanoes, nature's furnaces, to test ideas about how heat was transferred and how heat was related to temperature. It was also understood that a measurement of lava temperature would also reveal the temperature of the earth's interior. But, by the early 20th century, no one had yet measured the temperature of molten lava where it was spouting out of the ground.

An early crude attempt to get some idea of the temperature of molten lava was made by Scottish chemist Sir James Hall who, in the spring of 1785, was traveling through Italy when he heard that Vesuvius was pouring out lava. He hurried to the volcano. He managed to find a way close to a red-flowing river of molten rock, though still miles from where it was being erupted from the ground, and threw coins into it, watching to see which ones melted. The silver ones always did. The copper ones sometimes melted, an indication that the temperature of the lava flow close to where he was standing must be higher than the melting temperature of silver and close to that of copper. Years later, the melting temperature of copper would be determined to be about 2,000°F (about 1,100°C).

Yale professor James Dwight Dana made a remarkably good guess of the temperature of a lava fountain at Kilauea years after a visit to the volcano in 1840. About forty years later, he was standing close to an iron furnace and decided that the lava fountain he had seen was slightly redder, and, hence, cooler, than the "white-hot" slag being drained from the furnace. The iron slag was known to be about 2,400°F (about 1,300°C). And so Dana guessed the temperature of molten lava at Kilauea must be a few hundred degrees cooler. It was a good guess, about 2,200°F (1,200°C), and that value became the gospel and was reported in geology textbooks. And, yet, it was a guess.

As his first major experiment at Kilauea, Jaggar decided to settle the issue by measuring directly the temperature of molten lava spouted by Old Faithful. But how could he do it?

He enlisted the aid of Arthur Day, the director of the Geophysical Laboratory at the Carnegie Institute of Washington and one of the country's foremost authorities on high-temperature laboratory experiments, having determined the melting temperatures of many metals, including platinum, which has a melting temperature of 3,200°F (1,760°C), the highest melting temperature yet recorded for anything. Day had also measured the melting temperature of several different types of rocks, but all of his work had been done in a laboratory. He would have to design a device that could withstand the severe conditions of an erupting volcano.

Jaggar's initial idea was to use a kite to position the device, then lower it down and into Old Faithful, but Day dissuaded him from the idea. Any such device would be too heavy to be carried by a kite. They needed a cable stretched across the crater, then, by means of a traveling pulley, position and lower the device down to the lake. Jaggar had MIT civil engineers design the cable-pulley system. Day was in charge of designing the device that would measure the temperature.

After months of work, the device and the system were ready. On March 2, 1911, Jaggar wrote to Thurston, informing him that "an apparatus for spanning the crater with a cable and lowering an armored electric thermometer to the surface of the lava" was completed. Day would be sending Ernest Shepherd, one of the chemists at the Carnegie Institution, to oversee the measurements. Frank Perret had agreed to join Jaggar at Kilauea that summer, so there would be two men to work with Shepherd on the temperature experiments. Afterwards, Jaggar assured Thurston, the routine work of a volcano observatory would be started.

But personal complications forced a change in the plans. On April 24, Jaggar wrote again. He confirmed the arrival of Shepherd and Perret, but he would not be able to join them.

"I find that I, personally, shall be detained in New England owing to the condition of Mrs. Jaggar's health," he wrote.

What was her condition? She was three months pregnant.

———•———

Perret and Shepherd arrived in Hilo on July 2. Thurston met them and took them to breakfast. They then boarded the train. At the end of the train line, a motorcar from the Volcano House was waiting to take them the final ten miles, the stagecoach having been replaced a few months earlier.

Bishop Thomas Jaggar

Anna Lawrence Jaggar

May, Anna Louise and Thomas Jr. Jaggar

Helen Kline Jaggar and three-
month old Eliza, February 1912

Thomas Augustus Jaggar, Jr., 1916

Isabel Peyran Maydwell Jaggar,
undated photograph

Frank Alvord Perret, 1909

Perret using headphones attached to a megaphone to listen to steam emission near Naples, Italy. *Courtesy Library of Congress*

Perched lava lake within Halemau´ma´u, March 1894. *Courtesy Bishop Museum*

Hawaiian Volcano Observatory, 1913. *Courtesy Hawaiian Volcano Observatory*

Thomas Jaggar, Jr., in the Whitney Vault, located beneath the observatory building, October 31, 1916. The two upright metal poles on the left side of the photograph are the Bosch-Omori seismograph. *Courtesy Bishop Museum*

Volcano visitors descending into Halema´uma´u, Saturday, January 7, 1917. Isabel Jaggar is standing on the crater rim, second from left, wearing a white blouse. Lorrin Thurston is also standing on the crater rim, hands in his pockets. Alexander Lancaster is at lower right. *Courtesy University of Hawaii, archives; Photo by T.A. Jaggar*

Sounding the depth of the lava lake, January 23, 1917. Four men can be seen: one dark figure in the center of the photograph standing on a promontory and three figures, to the right, wearing white shirts. *Courtesy Hawaiian Volcano Observatory; Photo by J.J. Williams*

Two eruption columns rising to form a single large column over Mauna Loa, 8:30 A.M., May 19, 1916. Taken from the porch of the observatory. *Courtesy Hawaiian Volcano Observatory; Photo by H.O. Wood*

Aerial photograph of a lava flow of Mauna Loa headed for the fishing village of Hoʻopuloa, south Kona, April 11, 1926. *Courtesy Hawaii Volcanoes National Park*

Lava entering the sea at Hoʻopuloa, south Kona, at 6:21 A.M., April 18, 1926. *Courtesy Hawaiian Historical Society; Photo by Tai Sing Loo*

Ohiki, Jaggar's first amphibious vehicle, landing at a sandy beach in Kona, February 1928. *Courtesy Hawaii Volcanoes National Park*

Honukai, the second amphibious vehicle, on Alaska beach, summer 1928. Jaggar is at right. *Courtesy University of Hawaii; Photo by R.H. Stewart*

Thomas Jaggar driving *Honukai* in Hilo, October 18, 1940. Isabel Jaggar is seated in back, wearing a wide-brim hat. *Photo by J. Snedeker*

Thomas and Isabel Jaggar at the source of the 1919 eruption of Mauna Loa, October 1919. *Courtesy University of Hawaii, archives*

Ruy Finch, Isabel Jaggar and Thomas Jaggar sitting in front of the Volcano House, January 12, 1920. *Courtesy University of Hawaii, archives; Photo by L.H. Dangerfield*

Isabel and Thomas Jaggar, 1920.

Hotel manager Demosthenes Lycurgus was there to greet them.[*] They settled into a cottage owned by the hotel. Later the same day, Lycurgus sent them in his personal car to the edge of Halema'uma'u; a macadamized road to the edge of the crater had been completed the previous December. Volcano visitors could now ride in comfort to see molten lava.

Perret began his study of the volcano by walking completely around Halema'uma'u, stopping to take several photographs. He then watched the lake. The level of the lava was three hundred feet down from the crater rim, one hundred feet deeper than when Jaggar had seen it in 1909. Old Faithful was spouting at minute intervals.

Thurston arrived a week later to help with preparations for the measurement of lava temperature. Three carpenters from his lumber company came with him. He had the carpenters build wood anchorages on opposite sides of the crater that would be used to suspend the cable. While this work was underway, Shepherd checked the devices that, in turn, he would lower into the crater.

He had brought three devices of two different designs. Two of the devices were "resistance-type" instruments, meaning a temperature was determined by measuring the change in electrical resistance of a coil of platinum wire. The higher the resistance, the higher the temperature. The coil was insulated with a quartz tube that fit inside a nickel case that was placed inside a long iron box.

The other device, the "electro-element" instrument, had a sensor that consisted of two wire elements made of different metals. One element was of platinum and the other of a platinum alloy. When connected and one was kept cold while the other was heated, an electric current would run between the elements. The higher the difference in temperature between the two wire elements, the greater the electric current that would pass between them. The problem was keeping one element cold while the entire device was placed in

---

[*] George Lycurgus, the counterrevolutionary and owner of the Volcano House, was Demosthenes' uncle and was in Greece when Perret and Shepherd arrived at Kilauea.

molten lava. To solve the problem, the cold element was surrounded by a fifty-gallon steel tank that was filled with water, which added considerably to the weight of the entire device.

Each device, regardless of the design, weighed a few hundred pounds. Shepherd considered the two resistance-type instruments to be the more reliable ones, and so he decided to use these two for the initial tests.

After the cableway was completed and the traveling pulley was ready, Perret and Shepherd spent a few days hanging an instrument from the pulley and sending it out along the cable to the midpoint. At that position, they lowered the instrument a short distance, then pulled it back up and retrieved it. As each day passed, the cable-pulley system deteriorated. Strands of volcanic glass, Pele's hair, rising from the crater caught on the cable and fouled the wheels of the pulley. Acids in the volcanic steam ate into and weakened the guide wires that supported the wooden anchorages.

Finally, on July 20, they were ready for the first measurement. At least six people were needed to operate the cable-pulley system. On that morning, Thurston arrived with three guests from the Volcano House to help.

The first step, as Shepherd instructed the others, was to gather wood from the nearby forest for a bonfire. After the wood was gathered and brought to the crater's edge, a fire was started and Shepherd stoked the flames, waiting for the coals to be red hot. He then had Thurston and the hotel guests lift one of the resistance-type instruments and place it on the coals. They waited. After several minutes, Shepherd checked the needle on a meter attached to a pair of wires connected to the encased platinum coil. The indicator needle moved an expected amount. The electrical connections were good; the coil was responding to the high temperature. The fire and instrument were then doused with water. Once cooled, the instrument was lifted and attached to the pulley. It was then sent out over the crater.

A rope was used to control the sliding of the pulley down the cable. Shepherd had the pulley stopped when the instrument hung over the point where Old Faithful would erupt, about five hundred feet out from the crater rim. Old Faithful was then quiet. Next Perret operated a reel that, by use of a strong steel wire, lowered the instrument down into the crater.

No one thought to record how long it took the instrument to reach the surface of the lake, but, once it did, the tip of the two-hundred-pound iron-encased instrument broke through what looked to be a leathery crust and immersed itself halfway into red lava.

Shepherd, who had watched the slow descent, turned his attention to the meter. He pressed a switch. The needle did not move. He began to check the wire connections when he saw a small burst of lava shoot up around the instrument. Part of the fallback from the burst struck the steel wire and snapped it. Fifty feet of the wire and the instrument now lay on the crater floor. Those on the rim watched as the debris was soon caught up in the general circulation of the lake surface and rafted to one side where it was eventually pulled under, Shepherd recording the scene as "much as a fish carries the angler's line under a root."

The first experiment had been a failure. But they had two more instruments. Thurston said he had to go to Honolulu, but would return in nine days. They would try again then.

On the day before Thurston returned, a new crack opened on the floor of Halema'uma'u, sending a dense plume of smoke and gas fumes over the site where one of the anchorages was located and where the cable-pulley system was operated. It was now impossible to see the crater floor from this location. Nevertheless, when Thurston returned on July 29 and had three guests from the Volcano House with him, it was decided to make a second attempt immediately.

The second resistance-type instrument was used. Again, a fire was built to check the electrical connections. And, again, they seemed to work perfectly. When all was ready, the instrument was hung from the pulley and sent out over the crater, though, this time, to a place away from Old Faithful.

Because of the heavy smoke and fumes, Perret, who was operating the reel that lowered the instrument, could not see the crater floor. And so Thurston stationed himself at the other anchorage on the opposite side of the crater and had the two hotel guests at intermediate places on the crater rim between him and Perret. Thurston and the two guests were each given a flag. The idea was to use the flags to relay a signal from Thurston whether to lower or raise the instrument. Shepherd stationed himself near Perret, ready to read the meter.

It may have been confusion in understanding the flag signals or it may have been a gust of wind that caused the heavy instrument to swing wildly as it was lowered into the crater. Whatever the cause, as the instrument was lowered, it bumped against a rock jutting out from the edge of the lake. Shepherd checked the meter. The needle was not moving. The instrument was reeled back in.

As soon as the outer iron and the inner nickel cases were opened, Shepherd knew what had happened. Both the quartz shield and the platinum coil were shattered. The instrument was useless. Only the less reliable electro-element instrument was left to use. They would try once more the next day.

That evening Perret had an accident. He was returning by car to the crater. The car had stopped and Perret was getting out when the driver decided to move the car forward to a better position. Perret was thrown to the ground, striking his head on the hard lava. Someone who saw the accident was surprised that Perret had not broken his neck.

The next day, July 30, Thurston and Perret and Shepherd went to the crater. It was raining and three huddled under umbrellas to have a discussion.

The guide wires and the metal bolts that held the wooden anchorages together were deeply corroded. Both anchorages and the cableway would probably soon collapse and fall into the crater. The plume of smoke and fumes passing over one of the anchorages had intensified. Mats of Pele's hair now clung to the cable itself, so that soon it would not be possible to run the pulley out over the crater.

These dire conditions actually made the decision easy. Either they made a last try today or abandoned any future attempts. They decided today would be the day.

Thurston had managed to convince only two hotel guests to come to the crater's edge in the rain and assist. And so he enlisted his wife and two children to help him.

Because of his injury, Perret was unable to run the reels. He positioned himself on the opposite side of the crater and would send flag signals. The two hotel guests were given flags and stood an intermediate distances along the crater rim, ready to relay signals. Shepherd stayed close to where Thurston and his family would operate the reel, Shepherd finding a secure spot where he could watch the needle on the meter, knowing a reading may last for only a second.

Thurston, his wife Harriet and twelve-year-old son, Lorrin, had the most dangerous jobs. As Thurston cranked the reel, his wife would keep the wire tight on the drum while his son hustled back and forth, keeping the electrical wires from getting tangled. All three had to be on a narrow ledge between the anchorage and the crater rim. The crater is so large—a half-mile in diameter—that when coupled with the strong wind, the heat at the rim is not prohibitively strong. Meanwhile, his daughter, sixteen-year-old Margaret, would be watching for flag signals, calling out instructions to her father on how to move the instrument.

The rain was steady and Thurston heard a shout from his daughter to send the pulley and instrument down the cable. As the pulley and heavy instrument slid down, putting strain on the anchorage, every wooden strut and metal bolt creaked and groaned. The whole structure seemed ready to fall into the crater.

A second shout came from his daughter, and Thurston stopped the pulley. Then a third shout came. Thurston began to crank the reel, lowering the instrument. His wife and his son busied themselves at their assigned tasks. Another shout. And he stopped the cranking.

As the instrument was lowered, Perret could see it was positioned directly over Old Faithful, which was then quiet. At exactly 1:35 P.M., according to Perret's notes, the tip of the instrument touched red lava. At that same moment, Shepherd saw the needle rise steadily, then stop. Perret signaled to pull up the cable, then to lower it again. When the instrument touched a second time, a burst of steam shot out. The steel tank holding the fifty gallons of water needed to cool one of the sensors had ruptured. Just as that happened, a fountain of lava rose from Old Faithful, the down rush of molten rock snapping off the instrument.

"I regret that only one reading was possible," Shepherd wrote later, "but I have every confidence that it was a good reading."

The lava temperature measured that day was 1,850°F (1,010°C). And much has been made of this measurement—most of it critical.

First, the immersion of the instrument into the lake was too short a time for the platinum sensors to have felt the full heat of the lake, and so the temperature is certainly too low to be wholly accurate. Second, at these high temperatures, there should have been a chemical reaction between the iron casing and the platinum sensors, discounting the reliability of any measurement. Much more recent measurements, using a variety of techniques, indicate the temperature of a lava fountain at Kilauea is about 2,100°F (1,150°C), close to the guess made by Dana in the 1880s.

Nevertheless, what Perret, Shepherd and Thurston accomplished that day—and Jaggar should be included because he planned and organized the effort—cannot be discounted. It was a first. And it had been heroic. (Thurston made sure of this by publishing a detailed account of the effort, complete with photographs, on the front page of his newspaper.) No one had attempted anything like it anywhere in the world.

It set the cornerstone for more than a century of work at Kilauea and other volcanoes where measurements of lava temperature have been combined with microscopic examination of lava samples to relate, in the field, lava temperature and lava mineralogy, that is, to reveal the temperature conditions inside the earth and how various minerals are formed.

———•———

Shepherd left in August. Perret stayed two more months.

With material donated by the Volcano House and with the help of the hotel's handyman, Perret built a small hut on the rim of Halema'uma'u, the back wall literally a foot from the crater's edge. He had a large window with a glass pane installed on the back wall that quickly became hot to the touch and remained so for the three months Perret lived in the hut. The window provided a commanding view of the lava lake in all types of weather.

From this close vantage point, he kept a constant watch. He left only to gather groceries at the Volcano House or to use a photographic dark room that Lycurgus permitted him to assemble in a vacated storeroom at the hotel. It was within the hut that he wrote his weekly reports on the state of the lava lake, Thurston publishing each one in his newspaper. These reports are remarkable for their candor. For example, Perret describes how one day he managed to climb partway down the steep crater wall to where he found a narrow ledge. He followed the ledge to a cave where the hot walls were glowing a dull red. He photographed the cave and collected rocks. The next day, continuing the narrative in a matter-of-fact tone, he related how the ledge had collapsed during the night and was now inaccessible.

These reports—and his role in making the first measurement of the temperature of lava at Kilauea—made Perret a celebrity in the islands. Now when new guests arrived at the Volcano House, their first question was: Was it possible to meet Mr. Perret? And the answer was: Yes, it was. He was always somewhere within sight of the lava lake.

His hut consisted of four thin walls comprised of wooden slats nailed to a wooden frame and covered by a corrugated iron roof that corroded so quickly in the high concentration of sulfuric acid in the fumes that it gave little protection from frequent rains. The floor was hard lava rock. As soon as he moved in, Perret placed one of his "seismoscopes," as he called them, on the floor. It was a device he had invented consisting of a small weight attached to the top of a thin metal rod that was free to swing in any direction. Perret placed a large glass bell over the instrument so that the wind could not disturb it. As soon as he placed it on the rocky floor, the small weight and rod started to vibrate. It continued to vibrate for the entire two months he lived in the hut, sometimes more and sometimes barely perceptible.

In locating the hut, Perret insisted that it be built over a deep crack that he used for another experiment. He dropped a microphone connected to wires down the crack, then sealed the crack with concrete. He attached the headphones to the free-end of the wires. For hours, he would lie on a cot, the headphones to his ears and his eyes focused on the seismoscope, noting how the sound through the headphones or the swings of the seismoscope varied as the red glare of the lava lake filling his room rose or fell in intensity.

One person who visited the volcano that summer and met Perret—and there were hundreds of visitors—wrote to Jaggar about "the strange little man who seemed to thrive on air filled with the smell of sulfur."

"I was at the volcano last week," the visitor wrote to Jaggar, "and had several interviews with Perret. He is certainly an enthusiastic as well as a keenly intelligent worker. It is impossible to get him away from the little building he had erected at the brink of Kilauea. He is there night and day, and, from what I can gather, is so interested in gathering information concerning Pele's doings that he gets scarcely any sleep."

On September 17, after living at the crater's edge for fifty-five days and without giving warning to anyone, Perret packed his personal and professional things and moved into the Volcano House where he spent a few days to compile his notes. Then he left for Honolulu where he gave a public lecture about his adventures at Kilauea. Every seat in the thousand-seat auditorium was taken. People crowded in the aisles and in doorways, anxious to hear from the man who had lived at the volcano. As one person who attended remarked, "There were five times as many persons present last evening to hear about volcanoes as came last week to listen to the imminent danger of yellow fever in the community." And he electrified his audience, telling them of his experiences at Vesuvius in 1906 and his later adventures at the volcanoes of Stromboli and Etna and seeing the destruction caused by the Messina earthquake.

A few days later Perret met privately with Thurston who tried to convince him to stay and live at Kilauea and take charge of the scientific work. But Perret declined. He said he thought he could serve the study of volcanoes best by not devoting himself to any one place. With that final comment, he left and returned to Italy.

———•———

Thurston and Jaggar exchanged letters, discussing what should be done next at Kilauea. They agreed that someone should come and continue Perret's work as soon as possible. The obvious choice was Jaggar himself. In one of his letters, he explained why he could not come immediately.

"I am expecting an addition to my family about the end of October. This would make my going to Hawaii in November impossible, even if my work here would permit it, which it would not. I would have been with you in June were it not for this prospect which made my going to active volcanoes for the time out of the question."

Two weeks later, again writing to Thurston, Jaggar said he knew someone who might be willing to work at Kilauea permanently. "I have in mind a man, Mr. Ferguson, graduate of Harvard. . . . He is of an experimental turn of mind and might be a good man for superintendent of the observatory. . . . It is possible that I might get him to go . . . as I think he is in this country. I am sending out a letter today with a view to locating him."

Henry Ferguson had been one of Jaggar's students at Harvard. He had led the expedition to Iceland in 1905 when Jaggar had been unable to go and had almost joined Jaggar on the Alaska adventure. After graduation, Ferguson had taken a job as a government geologist in the Philippines. It did not go well. He had hoped to study the volcanoes and record the effects of earthquakes. Instead, as he wrote to Jaggar, "I have been troubled somewhat with dysentery and while well enough, I find it hard to keep up my enthusiasm." Ferguson was now back in the United States and working as a geologist in California, assaying gold mines. After his health problems in the Philippines, he wanted nothing more to do with a tropical climate, and that included Hawaii.

Jaggar asked others—George Hosmer, Reginald Daly, Norman Bowen—all associated with MIT and all men who would become experts in their fields, Hosmer as a topographer, Daly and Bowen as petrologists. None of them wanted to dedicate themselves to Kilauea. If one had, the subsequent story would have been different.

And so Jaggar wrote to Thurston. "Frankly, I know of no one interested who would do the work but Perret and myself. With Perret in Italy the question appears to narrow itself to me." Then he added, "Just as [Perret's] heart is in Naples, so mine is in Hawaii much more than it is in Boston."

Encouraged by Jaggar's latest letter, Thurston asked by telegram: "When will you arrive Honolulu?"

There was the practical concern of money, Jaggar answered. The MIT Executive Committee had refused to approve the spending of any more money at Kilauea, including that from the Whitney fund. And so Thurston took another step. His businesses were showing profits, as were those of many other prominent businessmen in the islands. On October 5, 1911, at a luncheon at the University Club of Honolulu, under Thurston's direction, a resolution was passed that established the Hawaiian Volcano Research Association. The primary purpose of the Association was to raise money from local contributors to support scientific work at Hawaiian volcanoes. In hindsight, this was the pivotal act that led to the establishment of a permanent volcano observatory, only the second of its kind in the entire world. But there was another hurdle to clear—Jaggar and his family.

His mother-in-law, Ella Kline, had arrived in Boston in October 1911 to assist her daughter in the birth of her next child. The Jaggars had already moved from the isolated stone mansion at the top of a hill in Brookline to a comfortable, though small, house on Concord Avenue in Cambridge, a few blocks from the Harvard campus.

Ella Russell Kline was a tiny woman. She was raised a Quaker and, after her marriage to George Kline, she joined the Episcopal Church, remaining a devoted member throughout her life. She was also the quintessential Victorian woman. Whenever she entered a room, even in her own house, she always walked erect and with poise and was properly attired wearing a dark skirt and light blouse that covered her neck, chest and arms and left only her hands exposed. When she finally approved of the installation of a telephone in the Kline house, she insisted that it be out of sight under the stairs and that a large mirror be hung nearby so that she could check appearance before talking on the telephone. She was also known to use a bath sheet that was pulled over a tub, leaving only one end open. After a maid had filled the tub with water, the maid would leave the room and Mrs. Kline would enter. She would disrobe alone, then slide into the tub and beneath the sheet and wash herself, only her head exposed.

On November 2, a daughter, Eliza Bowne, was born. Two weeks earlier, just about the time when his mother-in-law arrived, Jaggar asked MIT President Maclaurin and the Executive Committee to grant him a leave-of-absence to work at Kilauea. The request was granted.

The Jaggar family, including Ella Kline, spent Christmas Day at the rented house in Cambridge. The next day Jaggar said goodbye to his wife, his six-year-old son and newborn daughter. One can only imagined what glances, perhaps, what words, were exchanged between him and his mother-in-law as he left for the Hawaiian Islands.

CHAPTER NINE

A DREAM FULFILLED

I do not know just how long I will remain here," Jaggar announced to a waiting newspaper reporter as soon as he arrived in Honolulu. "I will probably stay here several months, but when I leave, I will leave the work in charge of an assistant, whom I expect to find about next week."

Finally, after a delay of more than two years, he was back in the Hawaiian Islands, landing in Honolulu on January 9. He spent a week in the city. His first task was to make contact with Thurston who took him to a private meeting with the territorial governor, Walter Frear, who asked Jaggar what he thought of Kilauea becoming a national park.

Jaggar favored the idea. And to make it into a reality, as well as to facilitate the planned scientific work, he suggested that a detailed topographic map be made of the summit region. And so Frear ordered it done. The governor sent word to the Territory's Chief Surveyor, Claude Birdseye, who was surveying a densely forested area on the steep north coast of the island of Hawaii. The governor gave Birdseye one day to move his men, their surveying instruments and their camping equipment to Kilauea and start preparing the needed map. In his rush to follow the governor's orders, Birdseye was careless and one of his mules kicked him in the face, causing him to lose a few teeth. To make matters more challenging, he planned to get married in Hilo the next Saturday. The wedding happened as planned, but a honeymoon was postponed so that the work at Kilauea could proceed.

Thurston was at Kilauea when Birdseye arrived and pointed out where the park boundaries would be. In all, the new national park would cover nearly sixty acres and include the crater Halema'uma'u. Notwithstanding his injuries, Birdseye, who happened to be a distant cousin of Clarence Birdseye of frozen-food fame, completed the survey in the remarkably short time of three months. Three additional months were required to prepare the actual map, which Frear carried to Washington, D.C., and showed to Congressional members the following fall. Another four years of lobbying were required before Congress approved the establishment of Hawaii National Park.

Jaggar arrived at Kilauea on January 17, 1912, reaching the Volcano House at noon. After lunch, he reviewed the recent activity recorded in the *Volcano House Register* and in several newspaper articles pasted into a scrapbook. According to one of the articles, the New Year's Eve party at the Volcano House had been interrupted when revelers standing on the hotel's verandah heard the roar of distant lava fountains and saw red glowing rocks flying above the rim of Halema'uma'u. Many of them raced to their cars, then sped down the road to the crater, some of the revelers hanging onto the outside of vehicles. The party resumed at the edge of the crater just before midnight. A few people found a way down to the edge of the lava, which, on this night, was only thirty feet below the rim; the highest level the lake had been in almost twenty years. During the next two weeks, the lake had quieted and the level had dropped. When Jaggar arrived, it was almost two hundred feet down.

At mid-afternoon, hotel manager Demosthenes Lycurgus provided Jaggar with a car and a driver to take him to the crater. He arrived at 4:30 P.M. Jaggar took notes: The lake was nearly circular. Two large craggy islands of solidified lava stood near the center of the lake, rafted around by the slowly churning lava. He sketched the islands and an outline of the lake. He paced out a baseline, then, with a handheld compass, he measured angles, later computing that the lake was 718 feet across at its widest point and 218 feet below the rim. Finally, he took out a pocket watch and timed successive bursts of Old Faithful. The average interval between bursts was thirty-six seconds. These notes were the beginning of continuous scientific observations of Kilauea that have continued ever since.

Such meticulous recording—Jaggar wrote in a perfect grade-school script—would be the hallmark of his work at Kilauea and at Mauna Loa when that volcano was erupting. With few exceptions, for the next twelve years, whenever he was at Kilauea and the lava lake was present, he made a point of standing on the rim of Halema'uma'u and recording the activity at least once a day, ideally twice a day if other work did not intervene.

And there was much else to be done. The owner of the Volcano House, George Lycurgus, made a special trip from Honolulu to meet Jaggar. In a letter Lycurgus had sent to Jaggar the previous month, he had promised the use of one of the hotel cottages to house scientific equipment. But so many guests had been staying at the hotel that Lycurgus suggested a new building be constructed. And so he took Jaggar to Hilo and introduced him to local business owners and convinced them to contribute money to the construction of the new building.

Thurston had also followed Jaggar to Kilauea. He selected the site for the new building, across the road from the Volcano House and close to the edge of a high cliff that formed part of the caldera rim. Whether Thurston knew this was a place important to local Hawaiians is unknown, but, to them, this short stretch of land with few trees and heavy brush and the occasional steaming crack was known as *Mauli-ola*, named for a local goddess whose healing powers were said to reside in the steam.

The ground, of course, had to be prepared before construction could begin. On February 6, in view of a small audience of hotel guests who followed Jaggar after breakfast to the site, the MIT professor took a machete and slashed through the brush, outlining the grounds of the observatory. Then three men, inmates on loan from the county prison—they had committed petty offenses ranging from gambling to drunkenness—finished the job of clearing the land. The use of prison labor was common in those years. Prisoners had built much of the road that Thurston had planned and that linked Hilo and the Volcano House. They had also been used to construct the new road that ran between the Volcano House and Halema'uma'u.

Nine days later Jaggar returned to the site. He took up four wooden stakes and marked out a square, eighteen feet on a side, using a compass to make sure the sides were aligned to the cardinal directions. Here a cellar would be dug to house scientific equipment. The alignment of the sides was necessary so that the equipment could be aligned in the same directions.

Again prisoners were used. They took a week to dig down through six feet of volcanic ash to hard rock on which the cellar would set. Atop the cellar would be a wooden building that housed the volcano observatory.

After three weeks, the building was practically done, so far as the carpenters and plumbers were concerned. It was built of Oregon pine. A wide porch wrapped around two sides of the building. The main entry was from the porch on the crater side. It opened into a well-lighted workroom where Jaggar planned to display the latest maps and charts that showed the progress of volcanic activity. Along two walls were a half-dozen small rooms. Most would be used for storage. One would be a small private office. Another, about the size of a closet, contained a kerosene stove to heat the building. A slightly larger room had been fitted for running water and would serve as a photographic darkroom.

As promised on the day he returned to the islands, Jaggar hired an assistant who would be in charge after he left and returned to Boston. The assistant was Francis Dodge, an athletic 22-year-old whose most recent accomplishment was the racing of automobiles along the few paved roads on the island, the Hilo newspapers often recounting his feats. Dodge's father, another Francis Dodge, was an engineer for Honolulu's Rapid Transit Company and had met Jaggar in Honolulu. The older Dodge hoped his son's passion for racing automobiles and other wild things would be tempered by the requirements of a regular job—and by the remoteness of Kilauea.

During this period of endless preparation, Jaggar took two days to make a partial ascent of Mauna, following an ancient Hawaiian trail to a point halfway up the volcano. He spent another day in Hilo ordering wooden furniture to be made for the observatory, including two large tables, a desk, chairs, a drafting table and a case of large drawers to hold maps. On February 17 he helped rescue a sailor from the USS *Colorado* who was visiting the volcano and who, in trying to take a photograph, had fallen into Halema'uma'u, landing on a ledge about halfway to the bottom. Remnants of the cable that Perret and Shepherd had used to measure lava temperature were used to rescue the sailor, who later died from a fractured skull. On another day, Jaggar stood close to the edge of Halema'uma'u and lowered a microphone—one of the new microphones designed for use with office equipment produced by the Dictaphone Company of New York—into several steaming cracks. He listened intently at each crack for sounds coming from the volcano. All he heard was wind passing through the fractured rocks. He spent an entire day walking around Halema'uma'u with another new piece of equipment, a Mansfield Water Finder, whose manufacturer guaranteed could detect "subterranean springs." The finder came in a handsome mahogany box with polished brass fittings. Inside was a large needle that could swing in one of two directions. Jaggar watched the needle as he circled the crater. Not once did the needle move.

And there was more to do. Camping equipment had to be purchased. Jaggar needed to meet local ranchers and judge who might rent horses and mules that were healthy enough to carry equipment to the summit of Mauna Loa. The ranchers invariably asked Jaggar what was the chance a future eruption of that volcano would send a lava flow over their land. Perret's hut had to be refurbished and repaired and moved; it had recently been vandalized. It was in the midst of such activity that a telegram arrived unexpectedly from Boston. It simply read: "The children are sick—Helen."

Jaggar left two days later, after he had conferred with Demosthenes Lycurgus who promised to watch over young Dodge and make sure he did not do anything reckless. Jaggar also wrote to Thurston, reminding him that his original plan had been to stay only a few months at Kilauea and to tell him why he had to return to Boston immediately.

He also reminded Thurston that Ernest Shepherd, who had supervised the lava-temperature measurements the previous summer, and Arthur Day, who had designed the lava-temperature instruments, would be arriving soon. They were coming to Kilauea to accomplish another scientific first: the collection of volcanic gases directly from the open vent of an erupting volcano.[*]

———•———

The role of gases in volcanic eruptions was self-evident to anyone who has seen an eruption. Lava froths furiously and fountains during eruptions, causing huge plumes of gas to rise up and fill the sky over volcanoes. Lacroix had concluded that these gases had been dissolved within molten rock, then released. But what types of gases were involved?

---

[*] Fortunately, when Jaggar returned to Boston, he found that the health of his children was not serious. Both had had whooping cough and his son had suffered from laryngitis, but both were soon in good health

Those with a sulfur component were an obvious choice. Sulfur is the stench common in volcanic regions, an odor that resembles that of rotting eggs. In the same region, if one looks into hollows or caves, one is apt to find scatterings of dead insects or the occasional dead mouse or small bird or, in extreme cases when conditions are right, the carcasses of large animals, as Jaggar had seen at Death Gulch at Yellowstone. Such deaths were caused by the presence of carbon dioxide that, because it is heavier than air, accumulates in hollows and caves. Many nights I have sat close to where molten lava is gushing out of the ground and have been intrigued by, among other things, the flickering of thousands of tiny pale blue flames. The pale blue flames indicate the burning of hydrogen gas. But which of these is the major component of the gases emitted in volcanic eruption?

In the 18th century, when the first systematic observations of volcanoes were being made, people rushed to the conclusion that different volcanoes emitted different types of gases. Etna was an emitter of sulfur dioxide. The perennial plume that hung over Vesuvius for most of the 18th and 19th centuries was thought to contain primarily hydrogen sulfide. The deaths of grazing animals in places such as the Dieng Plateau in Indonesia, an area surrounded by volcanoes, was thought to be caused by the sudden release of carbon dioxide.

In the 1860s, during a tour of Italian volcanoes, French chemist Ferdinand Fouqué tried to settle the question. Included in his equipment was a large watertight leather bag that he would hold up and collect as much gas from the edge of a volcanic crater as possible. He then examined the contents of the bag. Invariably, small beads of water had formed. And dissolved within the water were a variety of acids: hydrochloric acid, sulphuric acid and carbonic acid. From this simple experiment Fouqué concluded that different volcanoes probably did emit gases, primarily those that contained hydrogen, sulfur and carbon in varying combinations, but all of the volcanoes did have one gas in common: They all emitted water vapor.

That is where the measurement of volcanic gases stood for nearly forty years. Then, in 1902, Swiss chemist Albert Brun was inspired by the eruption of Mount Pelée to study volcanoes. He had an advantage over others, like Jaggar and Perret, who had also been inspired by that eruption: Brun had a personal fortune to finance his work and his travels.

He began with the volcanoes of southern Italy—Vesuvius, Etna and Stromboli. He would walk into a field of steam and smoke and insert one end of a long glass tube into a hot crack. When he thought the tube had filled with gas, he would remove the tube and put rubber stoppers at each end of the tube. He would then take the tube to a laboratory for analysis. He repeated this at the volcano Hekla in Iceland, on the volcanic islands of Tenerife and Lanzarote in the Canary Islands in the Atlantic Ocean and Reunion Island in the Indian Ocean. From there, he continued around the world to volcanoes in Indonesia, Japan and South America. His last stop was the Hawaiian Islands.

His travels took eight years. He reached Kilauea in August 1910. Here, he realized, was the best opportunity to collect volcanic gases because one could stand right at the edge of a crater that held molten lava. And so he took considerable time to devise a new and more elaborate sampling method and took considerable care in collecting his samples.

He constructed a 150-foot-long train of glass tubes connected by rubber joints. He dangled the train of tubes over the crater's edge. At the upper end he attached a small pump that he cranked to draw gases into the tubes. After several minutes of cranking, satisfied that enough gas had been drawn from the crater, he retrieved the train of tubes and analyzed the contents. He repeated the experiment at several places around the crater.

For every gas sample he retrieved at Kilauea and at the other volcanoes the result was the same: In none of the tubes was there any sign of condensed water vapor, as Fouqué had seen in his leather bags. Instead, Brun detected a wide variety of gases, mainly, hydrogen sulfide, sulfur dioxide, nitrogen and carbon dioxide. Brun made another discovery that he thought was important: There was always a thin layer of salts lining the inside of the tubes. But there was never any water.

In June 1911 Brun published a book, *Recherches sur l'exhalaison volcanique*, in which he detailed his work and his conclusions. He suggested the beads of water Fouqué had collected had come from the atmosphere, not the volcano itself. He also suggested the presence of salts and sulfur was the key to understanding volcanoes and supported the old idea, popular in the 18th century, that volcanic eruptions were powered by chemical explosions—and not the expansion of gases, as is known today.

One reviewer of the book decided that Brun should now "be ranked as one of the most remarkable of contemporary workers in science." His work had changed everything scientists thought they knew about volcanoes. The review ended with the comment: "Those interested in volcanoes should consult the book for themselves. It is to be devoutly hoped that before long its results will be critically tested in the field. Considering the peculiar facilities offered by Kilauea it is perhaps not too much to expect American scientists will soon take up the challenge."

Shepherd and Day had decided to do exactly that.

———•———

They arrived in Hilo on the morning of May 8, 1912. They were greeted by Thurston who took them to a luncheon held in their honor attended by several of Hilo's leading citizens. After eating and rising to say a few words in support of Jaggar and his plan to start a volcano observatory at Kilauea, Shepherd and Day boarded a special train—Thurston had held up its departure until the luncheon was finished—and headed for the volcano.

At the summit they introduced themselves to Frank Dodge, whom Jaggar had already directed to assist them in their efforts. The trio planned to descend into Halema'uma'u and collect gases directly from one of the numerous vents where lava was frothing and bursting up out of the ground. But Dodge was not yet familiar enough to lead the fieldwork, and so he hired volcano guide Alexander Lancaster to lead them into the crater.

Lancaster's background is difficult to trace. When asked by volcano visitors what was his origin, he gave a variety of answers. The more common ones were that he was Hawaiian or Negro or Cherokee Indian. He told his grandchildren that he was a lost prince from India. What little is actually known about him comes from a few documents.

Alexander Peter Lancaster was born on January 21, 1861, in Virginia. Given his dark complexion and the timing and the place of his birth, he was almost certainly born a slave. At age ten he had left the south and was attending public school in Oakland, California, probably the extent of his education. He was barely five feet high, weighed less than a hundred pounds, had perfect vision in both eyes and could hear slightly better in his right ear than in the left one.

On Christmas Day 1880 a small cargo ship, the *Eureka*, arrived in Honolulu from San Francisco. In addition to the thousands of sacks of flour and grain and the dozens of chickens and ducks on board, there were also thirteen passengers. One gave his name as "A. Lancaster." He is possibly the man who would later become the volcano guide.

The first mention of Lancaster at Kilauea is in an entry in the *Volcano House Register*. On August 21, 1888, the Reverend Charles M. Hyde who was visiting from Honolulu wrote: "At 3:30 P.M. started with Aleck Lancaster as guide." There are several other entries about him, all laudatory. One tells how he was kneading a lump of molten lava that he had just retrieved with a long wooden stick from the lava lake when a mass of lava seemed to surge toward him. He had his back to the lake. Those nearby started to scream. He turned and looked over his shoulder, determined the oncoming wave presented no danger, then calmly resumed his work. He was "the trusty volcano guide" and "a great favorite." And during the forty years he worked as a guide, he probably saw more volcanic activity than anyone in history.

On the day of the descent into Halema'uma'u, Dodge determined the lava level was 219 feet below the crater rim. The upper half of the crater wall was then nearly vertical with rocky overhangs. The lower half was a steep pile of loose rocks.

Lancaster began the descent by wedging a log into a crack near the crater's edge. He tied one end of a steel cable to the log and the other end to a long rope ladder. He then threw the rope ladder over the edge.

As he descended, Lancaster knocked away loose rocks with a hammer, stopping, when necessary, to untangle the ladder as he went. When he reached the bottom rung, he called for a second rope ladder—this one was rolled up—which Dodge lowered to him on a rope. Lancaster tied one end of the second ladder to the bottom of the first, tossed the second one down, and continued his descent. When he reached the top of the loose rock, he climbed back up to the rim.

He threw a long iron pipe and a wooden walking stick into the crater. Then Lancaster descended again, this time carrying a knapsack filled with twenty half-liter glass tubes. He also carried a hand-cranked suction pump made of iron. When he reached the bottom of the second ladder and was again standing atop the pile of loose rocks, he signaled for the next man to come down.

Shepherd was next. He also carried a knapsack. His was filled with notebooks and a camera. Then Arthur Day descended. He knew he would be unsteady, and so an Alpine belt was tied around his waist and a safety rope attached to the belt. Then, as Day descended, Dodge, still on the rim, held the free end of the safety rope while Shepherd guided him down. As soon as Day reached Shepherd and Lancaster, Dodge descended, also with a knapsack.

The four men now stood halfway down and inside the crater. Fumes were rising all around them. There was not yet any terrific heat, but breathing was difficult. Dodge reached into his knapsack and pulled out breathing equipment for each man: a rubber nose mask in which a water-soaked sponge could be held. Dodge sprinkled water from his canteen onto each sponge. He then handed a mask and sponge to each man, instructing him that, as the sponge absorbed gases, if the stench of noxious ones became too great, he was to pull out the sponge, squeeze it dry, then sprinkle more water and reapply the mask.

When everyone was ready, they continued their descent, each man crawling slowly backwards down the steep slope of loose rocks and choosing his own path, careful not to send a rock on a man below him. After an hour, all four were standing on the crater floor. The fumes were so thick that day that they could not see the crater rim.

Lancaster led the way, using the wooden walking stick to tamp out a safe path, occasionally, striding over glowing red cracks. The others followed.

He led them close to a small cone where small pieces of molten lava were being flung a few feet into the air. Once there, Shepherd gathered and assembled the equipment, attaching the suction pump to one end of the iron pipe. He attached an iron elbow joint to the other end. Then, with great difficulty, the four men swung the pipe out and maneuvered it so that the open end of the elbow joint slid into the top of the bursting lava cone. Shepherd then took one of the glass tubes and connected it to the exhaust end the pump. When all was ready, Lancaster crouched down and cranked the pump, pulling volcanic gases straight from the cone and through the pipe and the pump and into the glass tube. The others gathered around and stared at the tube.

Within minutes, a clear liquid appeared. Lancaster continued to crank. After ten minutes, a puddle of the liquid had accumulated inside the tube. A new tube replaced the one already used. Another sample of the clear liquid was obtained. Then a third tube was used and a fourth until all twenty tubes had been attached.

Of the twenty tubes, eight remained intact and without cracks as the pumping continued and after the four men had made it out of the crater. And all eight gave the same result: The clear liquid was water. This experiment, with improved equipment—Day and Shepherd would later use vacuum glass tubes and, instead of using a long iron pipe, would hold tubes directly over bursting lava cones—would be repeated at Kilauea and at other volcanoes, setting the standard for years to come concerning the collection of gases at volcanoes. The result has also been the same at every volcano: By far, the main component of gases emitted by volcanoes is water vapor followed by smaller, though significant, amounts of carbon dioxide and sulfur components and a very small amount of hydrogen.

After five hours on the crater floor, the four men made the return climb. The next day, on the front page of the *Pacific Commercial Advertiser*, Thurston heralded the experiment as "a hazardous undertaking" made by "determined men." Two days later, the crater floor dropped suddenly; some sections of it fell more than twenty feet. The place where the four men had stood was now a dark abyss filled with heavy smoke.

———·———

In 1912 the Hawaiian Islands were in the midst of their first boom in tourism. Full-page advertisements in national magazines and newspapers lured tourists with the promise of picture-perfect sandy beaches lined with coconut palms and an erupting volcano. A musical play, *Bird of Paradise*, opened on January 8 at the Daly Theater on Broadway in New York. The playwright was Richard Tully who had spent the summer of 1910 at the Volcano House where he had written the play. The play's popularity—the final scene was the sacrifice a young native woman to the volcano god Pele—increased interest in the islands as a tourist destination. In 1912 more than ten thousand people visited the Hawaiian Islands. And almost all of them came to Kilauea.

Thurston knew that to continue the appeal of Kilauea it was important to have a volcano observatory, an attraction the visitors would enjoy. And so, though Jaggar had been back in Boston for only one month, Thurston wrote to him asking when he might return.

Jaggar wrote back saying that he had met with Maclaurin and that they had discussed "whether it will be wise for me to take charge of the Kilauea station for a term of years." Maclaurin said no. There was administrative work for Jaggar to do and students to teach and whose work needed to be supervised. Also there was the issue of money. Maclaurin was adamant: No more money from the Whitney fund could be used in Hawaii.

Thurston, who, as a businessman, was used to negotiations, made an offer. He would finance all volcano work, except Jaggar's salary, which would be paid by MIT. Furthermore, MIT would have "scientific" control of the volcano observatory, the staff at Bishop Museum in Honolulu would have "local" control and Thurston would retain "administrative" control. And the new institution would be known as "Bishop Museum's Center for Volcanology."

Surprisingly, Maclaurin did not dismiss the offer out-of-hand. A month of negotiations followed, a flurry of letters, memos and cablegrams exchanged among Thurston, Maclaurin and Jaggar. Finally, Thurston conceded all points. He would supply all the money, which included paying Jaggar's salary, for a period of five years, some of the money to come from subscriptions from members of the Hawaiian Volcano Research Association. Bishop Museum was eliminated.* MIT would have "full" control of the observatory. And Jaggar would be given a five-year leave-of-absence to start the observatory and be its first director. The new institution would be known as "The Massachusetts Institute of Technology's Hawaiian Volcano Observatory." And, finally, in recognition of her financial contributions to MIT, the concrete cellar under the observatory building where scientific instruments would be housed would be named for Anne Whitney and would be known as "the Whitney vault."

———•———

On the eve of his departure from Boston—Helen and the children would come to Kilauea later—Jaggar was honored at a farewell

---

* Actually, Bishop Museum had tried to start its own volcano observatory between Perret's departure in October 1911 and Jaggar's return in January 1912.

dinner given by his MIT colleagues, including Maclaurin. After the last dinner course was finished and cigars had been passed around, the conversation turned to the risk of giving up a secure job in Boston. It then proceeded to the terror of sailing distant oceans and of living in foreign countries. And, finally, the men talked of the danger of volcanoes and of exposing their families to the awfulness of leprosy and other diseases common in the tropics. After the last topic was discussed, Jaggar, hoping to lighten the mood, rose and addressed his colleagues who he would soon leave. Yes, he acknowledged, such challenges exist and he would confront each one if it turned up, but, for now, finally sparking a jovial tone, "I ain't dead yet!"

On his way to the Hawaiian Islands, Jaggar stopped at the University of California in Berkeley to collect the man who would be his assistant, Harry Oscar Wood. Wood was a sour-faced man who often complained of catarrhal colds and other recurring ailments. But he was a capable man and one of the few, in this era, who knew anything about science of earthquakes.

Wood had been one of Jaggar's students at Harvard. After graduation, Wood was hired as an instructor in geology at the University of California where he was put in charge of teaching a laboratory class in mineralogy. In the hours immediately after the 1906 San Francisco earthquake, which had occurred in the early morning, Wood went out and began compiling information for a map that would show the intensity of damage produced by the earthquake. A month later, he was placed in charge of two seismographs owned by the University. Both were Bosch-Omori seismographs.

But, as the years passed after the earthquake, Wood was not happy at the university. He felt overworked and underpaid. The latter was certainly true. In the eight years he had worked at the university, he had never been promoted or received a raise in pay. And so, when Jaggar asked him about working at Kilauea, he was eager to leave.

Jaggar and Wood arrived at the volcano on June 13. On the first day of the next month, July 1, 1912, Thurston started to pay their salaries and other bills accumulated by the volcano work. That day is the official starting date of the Hawaiian Volcano Observatory.

Jaggar quickly established a daily routine. Work began at 8 A.M. with a trip for either him or Wood or both to Halema'uma'u where the lake level was measured and important changes since the previous day were noted. Then back to the observatory building where measurements of air temperature and humidity were measured and wind speed and direction and cloud conditions were noted. At around midday, a careful measurement was made using a solar transit to determine the exact moment of local noon. From that determination, the observatory's two pendulum clocks were reset. Finally, a wide-angle photograph was taken from the observatory of the summit region of Kilauea and the distant profile of Mauna Loa.

In the afternoon, Wood worked on setting up and adjusting seismographs in the Whitney Vault. Jaggar answered correspondence and met with visitors, explaining the purpose of the observatory. During the first few months after the observatory was established, he took a few trips, each one lasting a few days, to familiarize himself with the location of recent lava flows of both Kilauea and Mauna Loa.

The question soon arose as to the future of Frank Dodge. In fact, Jaggar had made plans for him. He had arranged for Dodge to enter MIT in the fall and prepare himself to be a volcano observer. "The idea wasn't so screwy if placed before the right boy," Dodge would later write. But it was not an idea that "heated" him up. Instead, he "just got lukewarm. Just warm enough to say yes."

His father gave him $400 for the trip to Boston. Before he left he made a list of women's addresses in San Francisco he copied from the Volcano House registry.

Arriving in San Francisco, he spent two weeks in the city playing poker, going to bars and dating women. One morning, after waking up from a stupor, he found that he had only $35 left in his pocket. "There went any chance of college," he remembered of the moment. Unwilling to face his father, he drifted around, finding work wherever he could. He worked at land surveying, then as a boatman on the Colorado River. In old age, he settled down and maintained the stream gauge at Lee's Ferry at the upper end of the Grand Canyon. He never did return to volcanoes.

To take care of camping equipment and to be a field assistant, Jaggar hired Alexander Lancaster. His pay was $25 a month. Lancaster had a tendency to drink, and so Jaggar paid him small amounts through a month until it totaled $15. On the last day of the month, his wife, Makaweo, came to the observatory and Jaggar paid her the balance.

A third person was hired the first year the observatory existed. She was Emily Farley. In view of what soon happened, it is important to understand who she was.

Farley had sailed on the same ship that had brought Jaggar and Wood to the islands in June 1912. In fact, Jaggar and Farley had met on the first day at sea, he writing in a notebook how he could contact her again once they had reached the islands.

Farley was twenty-seven years old. She was a recent graduate of Vassar College where she had been president of the French Club. She had acted in amateur stage productions, the story of "Pyramus and Thisbe" taken from Shakespeare's *A Midsummer Night's Dream* being one of them. After her graduation from Vassar, she had taught high school for two years in Newton, Massachusetts. Born in Boston, her mother was part Hawaiian and her father was a graduate of MIT. The Farley family had a substantial amount of wealth as owners of a major mercantile business in Boston.

When she met Jaggar, Farley was on her way to visit relatives in the Hawaiian Islands, where she planned to stay an undetermined length of time. In fact, she visited the Volcano House for a few weeks in July 1912, soon after arriving in the islands. On October 1, Jaggar hired her as his receptionist and secretary.

In November the two of them traveled together to Honolulu to attend meetings with Thurston and other members of the Hawaiian Volcano Research Association. Emily Farley left Honolulu on the morning of November 26 and sailed back to Hilo. That afternoon Helen Jaggar and the children arrived in Honolulu.

CHAPTER TEN

A LOVE LOST

B y the time Helen Jaggar left Boston, both the stone mansion at Brookline and the seaside cottage at Lewis Bay in
West Yarmouth had been sold. So had the Peerless Touring
Car. Most of the wedding presents and household furnishing had
been auctioned off. In September 1912 Helen and the children
traveled by train to San Francisco. The Pullman car they rode
in was filled with forty-one pieces of luggage, a baby's hamper,
a crib and two refrigerated boxes of milk. While in San Francisco, they stayed with her parents. On the eve of their scheduled departure, she became mysteriously ill and was taken to a
hospital.

Six weeks of recuperation followed. Finally, on November 19,
she and the children sailed, arriving in Honolulu seven days later.
Her husband was on the dock to greet them.

The Jaggar family, now reunited, stayed in Honolulu for two days, then left for Hilo. As their ship departed the harbor, her husband pointed to another ship and said that it would soon be sailing on an expedition to the South Pacific and that he expected to be on it. Helen said she was surprised that he had allowed her to leave San Francisco without telling her about the expedition and his departure. She protested vigorously about being left alone so soon after her arrival. Well, he said, she could return to Honolulu. That was nearly as bad, she answered, because she did not know anyone in Honolulu except for a few acquaintances she had just met on the ship. The possibility of him joining the expedition was discussed again over the next few days. In the end, the ship sailed without him.

The Jaggar family settled into a cottage owned by the Volcano House and located a short walking distance from the hotel. It consisted of two small rooms. The front one served as a parlor and was barely large enough for two chairs. The larger one was the bedroom and came furnished with a dresser, a small table and a bed. The baby's crib would fill most of the remainder of the floor space.

They took their meals at the Volcano House restaurant. Helen now saw the hotel as "a roughish place" where "the rooms were bare to barrenness, the food execrable, and the milk supply the worst [she had] ever seen."

There was a continuous procession of people who came to stay at the hotel, but few of them interested her. "I was not at all happy," she would remember of this time. Before she was married, she was living in a mansion in San Francisco, a city of dazzling lights and endless shops and where there was always a party to attend. She had given that up for Boston and its gloomy weather and snowy winters. At least in Boston one could attend a stage play or listen to an orchestra performance. But, here, on an island in the Pacific, at the top of a volcano where it rained frequently and where every night was dark and cold, where she was miles from any town of any consequence and was surrounded by a dense forest, where there was little or no social life and no stage or music performances—this is where she now lived. And her life soon became worse.

For the past few years, her father-in-law, Bishop Jaggar, had been living in southern France serving in the mostly honorary position as head of the American Protestant Churches of Europe. On December 12, 1912, he died. The news arrived in Hawaii three days later. His son then did an odd thing. The next morning he abruptly left Kilauea and sailed to Honolulu, arriving late the following afternoon. He stayed one night and sailed the next morning back to Hilo. What did he do in Honolulu for one night? From the brief account left by Helen Jaggar, he probably went to see Emily Farley.

Thomas Jaggar was back in Hilo on December 20. He and Helen spent the night at the Hilo Hotel. The next morning they were out shopping for Christmas presents for their children when they got caught in a torrential rain. As Helen remembered the day, "We had been out driving, and as soon as we got out of the carriage near the hotel the rain pelted on the pavement so hard that it bounced up almost to our knees. We were drenched before we got inside, but that was just the beginning."

The rain now fell harder, making it impossible to see the other wing of the hotel a hundred feet away. Then the rain stopped as quickly as it had come. "In that hot climate so much water is distinctly unpleasant," wrote Helen, "for as soon as the downpour ends, everything steams. And believe me, we steamed that day."

As soon as the rain ended, her husband said he had something to tell her. He said that Emily Farley would be arriving that day. What he said next Helen refused to ever reveal, except to write, "From the moment he finished the story I would have nothing more to do with Mr. Jaggar than allow him in the house."

Three days later was Christmas Eve. The Jaggars spent it in their cottage at the Volcano House. A native ohi'a tree with bright red blossoms was their Christmas tree. The children received gifts the next morning in their stockings.

The next day the Jaggar family sailed for Honolulu. Helen and the children settled themselves in a hotel room. Her husband stayed one night. He returned to Hilo and resumed the work of the observatory.

———•———

Exactly a year earlier, the surface of the lava lake had been within thirty feet of the crater rim and the lake had covered more than two hundred acres. Now it was four hundred feet down and covered less than one hundred acres. Over the next few months, it continued to drop and diminish in size.

On January 7 Jaggar began making hourly measurements of the dimensions and the level of the lake. He did so for twenty-two consecutive hours, forced to end his effort when heavy rain fell, which created fumes that were so thick that he could no longer see the bottom of the crater. He returned the next day and made an additional twenty hours of measurements, again retreating during a heavy rain.

On the evening of January 18 he was standing at the edge of Halema'uma'u when another rainstorm struck, this one more intense than the previous two. This time he stayed, curious to see what would happen as the cold rain chilled the lake. During the first half-hour of heavy rain, the rate of churning of the lake noticeably slowed. Eventually, it stopped. He remained at his post. An hour passed, then two. Slowly lava started to fountain at one end of the lake, splashing molten rock high on the crater wall. After several minutes, that activity stopped and the scene was again dark. He could hear sputtering and hissing. Then, without any forewarning, a huge lava fountain broke out in the middle of the lake and lit up the entire sky. The entire surface of the lake started to churn again. All the while, heavy rain continued to fall.

A month later he had a minor mishap at the crater. He was alone and busy making measurements when a wind came up and blew his notes for the last four days into the crater. He knew that one of the long rope ladders used by Lancaster and others to descend into the crater was stored nearby in a crevasse. He found the crevasse and retrieved the rope. He used it to descend as far as he could. Standing on the bottom rung of the ladder, he could see his notes about ten feet away. He stepped onto a narrow ledge and inched his way closer to the notes. When he had gone as far as he could, he used a long stick that he had brought with him and had prepared with a safety pin tacked to one end to retrieve his valuable notes.

By mid-March the lake level was five hundred feet down. Now there was only a small pool of lava barely visible from the crater rim. The lake of lava that had originally attracted him to Kilauea had drained away slowly. On April 1 Emily Farley resigned. Six days later, molten lava was last seen in Halema'uma'u. The next day Jaggar left for Honolulu. He had decided to reconcile with his wife.

———•———

Helen had moved into a small house where, as she would write, she and her children "might have been very comfortable if I had not been so distressed." Her husband visited, but to no avail. He could not persuade her to return to Kilauea. Instead, after another month, continuing to despair over her situation, Helen sent a telegram to the only person she knew she could rely on—her father.

George Kline arrived in Honolulu ten days later. Helen was thrilled to see him. But the thrill did not last.

Her father reminded her that he had tried to prevent the marriage, but Helen had refused to call it off. Now she had two children. The possibility of divorce, he said, should not be considered. It was her duty to follow her husband wherever he chose to live and no matter what he decided to do.

For two weeks Helen argued with her father and her husband. One afternoon she met with them and two lawyers. All four men tried to dissuade her from leaving the islands and returning to San Francisco. But she was determined. Finally, her father gave in. On May 16 she had one last private meeting with her husband. Later that day, Helen and her children and her father departed for San Francisco. Her husband returned to the volcano.

When the ship arrived in San Francisco, Helen's mother, her sister and two brothers were waiting on the dock. It was a cold reception and a solemn trip home. Once back, her mother informed Helen that the cook had been dismissed and that Helen would be taking her place. Furthermore, her six-year-old son would be restricted to a room in the attic whenever his grandmother was in the house. "He looks like his father," Ella Kline said. "I don't want him near me." As for the baby girl, she could sleep in a crib in Helen's room on the second floor.

For a year there were repeated shouting matches between Helen and her brothers. Helen cried a lot. Finally, her father secreted money to her and she and the two children moved to a house south of the city. A divorce was filed on July 3, 1914, and finalized a year later. Thomas Jaggar seldom saw his children again.

———•———

Though molten lava was no longer visible in the crater after April 1913, Jaggar was not idle after he returned to the volcano. The concrete walls and floor of the Whitney vault leaked during heavy rains, and so all of the scientific equipment had to be disassembled and removed and the walls and floor sealed and repainted. There was also an endless parade of visitors. Most were just curious. Others Jaggar had invited because they seemed to have something unusual to offer in the study of volcanoes.

One of those who Jaggar invited was Jehu Frederick Haworth of Edgeworth, Pennsylvania, who had made his money in the grocery business and who used it to develop larger and larger kites that could be used to take aerial photographs. Haworth had already done aerial surveys of small sections of two cities, Pittsburgh and Baltimore. He came to Kilauea in the spring of 1913 to photograph Kilauea.

Haworth arrived with a giant kite that he had designed specifically for use at the volcano. It had an aluminum frame covered with silk. It measured eleven feet in length and nine feet in width and was four feet tall. In a strong wind, Haworth estimated it could lift 160 pounds. It was tethered from a giant reel that held five miles of piano wire, which, when completely unwound, allowed a kite to fly as high as a mile above the ground. The reel was set in a large wooden cradle. It was turned by the operation of a small gasoline engine.

Haworth and Jaggar chose a site on a small flat of land between the Volcano House and the observatory building to set up the reel and launch the kite. It took two hours to fly it with a camera attached to the desired height. Haworth triggered the shutter by an ingenious device that he had invented and called a "messenger." It resembled a small umbrella on rollers. When all was ready, the "messenger" would catch the wind and run up the piano wire and trip the shutter. Then it took two more hours to reel the kite back in.

Haworth took more than a hundred aerial photographs of Kilauea, some showing great detail. Jaggar sent them to George Curtis who he had met when they had sailed to Mount Pelée and whose specialty was making clay models of American cities. Curtis made two models of Kilauea that were on display for years, both in museums, one at Harvard and the other in Washington, D.C.

The year 1913 ended with a prolonged rainstorm. No molten lava had been seen since early April, and, yet, he continued to make a daily trip to stand at the edge of Halema'uma'u, noting whatever changes, no matter how subtle, had occurred. On New Year's Day, he noted the crater was shaped like a funnel, the bottom more than six hundred feet down. The previous night a slab of crusted lava had slid down from the wall and was now covering part of the crater floor. He could hear a faint hissing. Puffs of steam were rising from a small hole in the floor. He timed the intervals between puffs and determined they varied from twenty to forty seconds.

He continued to make a daily trek to the crater, even though the first week of the New Year was windy and cold. And so was the second week. On January 13 the rain increased and seemed to drown everything. Now even the crater floor was not visible because of the great volume of fog the inclement weather had added to the normally thick volcanic fumes. On that day, again standing at the edge of

Halema'uma'u, he wrote in his notes that he could not imagine a more miserable condition.

That evening, back at the Volcano House, he saw a newspaper headline: The volcano Sakurajima in Japan had exploded. More than twenty thousand people were fleeing for their lives. Jaggar left the next day for Japan.

———•———

Sakurajima is an island that sits at the entrance to Kagoshima Bay at the southern end of Kyushu, one of the main islands of Japan. The volcano is almost identical in form and in size to Vesuvius. It exploded frequently during the 18th century, though most were minor ones. Since 1799 the volcano had been quiet.

That is, until early January 1914 when the people who lived on Sakurajima thought they were feeling slight shakings. They contacted local government officials, fearing an eruption was imminent. The officials contacted local scientists who examined the records from the few seismographs that then existed in Japan and decided that the shaking was not coming from Sakurajima but from other volcanoes that had erupted recently. The scientists announced that there was no immediate concern about Sakurajima.

At 3:41 A.M. on January 11 the ground shook so violently that people were awakened. A dozen more strong shakings occurred later the same day. The people of Sakurajima wasted no time; they gathered what possessions they could carry and headed for the shoreline where they used whatever means were available—fishing boats, sampans, makeshift rafts—to ferry themselves off the island and across the quarter-mile-wide strait that separated the island from the mainland. The exodus continued the entire day and throughout the next night. At 10:05 the next morning the first explosion rocked the island, an outburst on the west side. Ten minutes later there was a second explosion, this one of the east side where the exodus was still underway. Explosions continued from both sides, building in intensity. The climax came early in the evening when a fountain of fire rose several hundred feet above the west side of the volcano. From the base of the fountain poured a vast Niagara of fire, the brilliant streams of molten rock flowing quickly through orchards of orange and cherry trees and over terraced fields of sugar cane and into the sea.

The eruption was still in progress, though much diminished, when Jaggar arrived on February 3. The activity on the west side had ended. The eruption on the east side was still going, three streams of lava pouring into the sea. By the time he arrived, the strait of water between the island and the mainland was already filled. Where the sea had been as much as two hundred feet before the eruption, there now stood a lava flow as much as three hundred feet high.

He hired a local man who led him up the side of the volcano. Partway up the guide lost his way, the two men blinded by a heavy shower of ash and cinders. Eventually, the air cleared and the guide and Jaggar continued, reaching to within a few hundred yards of the site where lava was gushing out of the ground and flowing toward the sea.

A few days later, Jaggar and several others rowed out in a small boat to get a close look at the place where lava was entering the sea. As the small boat made its way, Jaggar trailed a thermometer in the water. When the boat was over the edge of the submarine flow and Jaggar and the others could see red incandescent rocks directly beneath them, he checked the thermometer. The temperature was 212°F (100°C), the boiling point of water. Small bubbles and steam were rising all around them. They continued, the boat crossing over the flow. Jaggar continued to check the thermometer. It rose to 240°F, then 260°F. The highest temperature he measured was 280°F (138°C), and then it fell. "It was a singular experience," he wrote. "With steaming water all around us, we had the unpleasant thought that if we should capsize we would be cooked." After returning to land, he walked along a beach, finding the carcasses of boiled horses and cattle and seeing thousands of dead fish floating in the sea.

Jaggar recounted the details of this and other adventures at Sakurajima in an article he wrote for *National Geographic* magazine. In the same article, he did what he probably considered his professional duty and praised the Japanese government for the quick response in organizing the evacuation of the island of Sakurajima. The success, he wrote, was owed "to good luck . . . to the instinct of people, to the wisdom of government, and to scientific societies." As already seen, the praise was not truly deserved.

The government had not provided warning nor could it have responded quickly enough to organize the evacuation. Instead, the exodus had been haphazard and, after the initial explosive eruptions on January 12, thousands of people were still on the island. More than a dozen died. After the disaster was over, the people of Sakurajima erected a memorial to those who had perished, one that is still known as "the distrust monument." It is a warning to future generations and reads, in part: "It is essential that residents put no trust in theory, but make preparations to evacuate immediately after detecting abnormalities."

Clearly, major steps needed to be made toward the prediction of eruptions.

Jaggar returned to Kilauea on April 21. Eleven days later, on May 2, for the first time in more than a year, molten lava returned to the bottom of Halema'uma'u. It was a pool about twenty-five feet across over the spot where Old Faithful had spouted. Over the next several months, the pool of lava grew slowly in size. Jaggar resumed his daily trips to the crater. Occasionally, he noticed a jet of gas throwing out bits of incandescent material in one corner of the crater floor.

As this minor activity continued, Jaggar and Wood made modifications to the Bosch-Omori seismograph. They replaced the original silk threads that suspended the heavy weights with short strands of piano wire retrieved from Haworth's kite experiments. They also invented a way to reduce the constant jitter of the long horizontal arms by adding small damping pots filled with cooking oil. The improvements soon paid off.

Shortly after noon on November 25, Wood called Jaggar down to the vault. The stylus on the Bosch-Omori seismograph was vibrating back and forth, tracing out a series of small earthquakes. They watched as ten such shakings were recorded over the next hour. All were too small for them to feel. Jaggar noticed the tracings were different in form from any that he had been able to associate with locally felt earthquakes or with rockfalls at Halema'uma'u. Wood suggested that something was happening at Mauna Loa.

The observatory was surrounded by fog all that afternoon, and so there was no view of Mauna Loa. Wood left the vault and the observatory at 6:15 that evening and was walking back to the Volcano House. The fog had started to lift. He stood and watched. After a few minutes, now with a clear view, he could see a single column of white fume rising from the summit of Mauna Loa twenty-five miles away. He waited. After several minutes, a second column rose. The two merged to form a broad anvil-shaped cloud, one that was clearly different from normal weather clouds. Now there was no doubt: Mauna Loa had started to erupt.

This was a first in volcano studies. Never before had anyone detected a flurry of earthquakes immediately before an eruption based solely on the use of a seismograph. Yes, many eruptions around the world had been preceded by earthquakes that were strong enough to be *felt*, but there were also many eruptions that seemed to begin without any seismic forewarning, including the most recent eruption of Mauna Loa in 1907. And so this was a milestone. Jaggar and Wood had shown that seismographs could give reliable and useful recordings (something that was still questioned after the Sakurajima eruption). And Jaggar would demonstrate this again and again for eruptions at Mauna Loa and Kilauea.

Today seismic recordings are at the heart of predicting eruptions. Time and again, a sudden flurry of earthquakes has proven to be the most reliable indicator of an imminent eruption.

———•———

Throughout the evening of November 25, 1914, every guest and every worker at the Volcano House came outside to stand and look toward Mauna Loa where the sky was lit with a bright orange hue. People as far away as Honolulu would report seeing the glow.

Jaggar and Lancaster, however, busied themselves that evening at the observatory, making preparations for a hasty trip to the summit of Mauna Loa. Jaggar enlisted the aid of a local man, nineteen-year-old John Gouvea, to assist them.

Jaggar had been to the summit of Mauna Loa only once before, in September 1913. And everything had gone perfectly. He and Lancaster had made it to the summit of the 14,000-foot-high volcano in one day, using horses and following an ancient Hawaiian trail on the southeast side of the mountain. The weather had been clear and the wind was calm. He and Lancaster had spent four pleasant days at the summit. He expected it would be the same this time.

On the afternoon of the second day of the eruption, after all preparations were completed, Gouvea, using one of the hotel's automobiles, drove Jaggar and Lancaster to nearby ranches where Jaggar attempted to secure riding and pack animals, but none of the ranchers would rent animals at any price fearing they would be returned exhausted and with their legs cut by fresh lava. And so an entire day was lost. That night Gouvea drove Jaggar and Lancaster to the west side of the island where they found lodging at a boarding house.

The next day was yet another day of frustration. Not until mid-afternoon was a rancher found who would provide animals, though the ones he did provide were of questionable health. In the end, Jaggar secured the use of three horses and five mules. He also convinced one of the ranch hands, Charlie Kaa, who was familiar with a trail that led to the summit of Mauna Loa from the west side, to guide them.

It would be a thirty-mile trek, mostly over open land. Kaa, Lancaster and Gouvea rode horses. Jaggar chose one of the mules. Each man led a pack animal. They had gotten a late start, and so they had managed to travel only a few miles when darkness forced them to stop and camp for the night. The weather was clear. The night sky over Mauna Loa was lit brilliantly by the eruption.

The four men were awake and were on their way before sunrise. They rode all day, without stopping for rest or to eat, Kaa repeatedly saying that they could reach the summit and see the lava fountains before sundown. The first half of the morning they traveled across grassland and through an occasional small tree grove. By late morning the trail was barren hard lava.

All went well until about 4 P.M. when a storm of sleet and wind came up, slowing their progress. Two hours later, just as the sun was setting, Jaggar thought he had the eruption in sight, but neither man nor beast could withstand the bitter cold and strong wind anymore. Ice had formed on the long hair and on the lips of the animals. The men were suffering from the altitude and from hunger. And so, less than a mile from completing their journey, Jaggar decided they should turn back. They retraced their steps, traveling back about a mile to a lower elevation, finally stopping when the sleet changed to rain. Here they spent the second night camped on hard rock.

Though everyone was exhausted, as Jaggar recounted it, Lancaster stirred himself to action and "was right on the job and self-sacrificing." He set up the small teepee-style tent that would be their only shelter, a large oilskin sheet held up by a single upright pole. He carried stones and piled them around the outside of the tent to keep the edges from flapping in what was an increasingly strong wind. After the animals had been unloaded and fed and the men were inside the tent, Lancaster lighted a small kerosene stove and warmed a mixture of canned milk and coffee and gave a portion to each man. And it was Lancaster who, throughout the night, left the tent occasionally to check on the animals and made sure they did not stray too far.

Jaggar, Lancaster and Kaa had each brought a change of clothes and they put on their dry ones as soon as the tent was ready. But young Gouvea had brought only the clothes he was wearing. He was shivering and cold and laid down in the center of the tent. Lancaster lay next to him, keeping the young man from freezing by using the warmth of his own body.

The intensity of the storm rose and fell. During a long lull, the men thought it might be over, but then came a sudden howl of wind and the pelting of increased rain. When that happened, everyone became anxious, wondering if the tent might blow down or be torn to shreds, leaving them exposed. "If we had lost the tent," Jaggar confided in his journal, "we could have been soaked over again and that might have been fatal."

Jaggar spent the night in a half-crouched, half-sitting position, buttressing his back against his small pack in an effort to keep his head higher than his legs, dealing, as best as he could, with an uneven rocky surface under him. Between pelts of rain, he could hear the animals tramp around the tent. Every now and then, he heard a squeal when two of the animals had a tiff. During those rare moments when both the wind and the rain stopped, he could hear the faint rumbling of lava fountains about two miles away.

The storm ended by daybreak. The sky was still overcast and the ground was covered with snow. The men could see clear weather far to the west. Jaggar looked over the men. Everyone was still tired and hungry. And so he decided to abandon camp and leave the equipment—Kaa would be sent to retrieve the tent and other items a week later—and return to the coast. They saddled the horses and mules—the seat and pommel of Jaggar's saddle had been chewed away by one of the animals during the night, making his ride difficult—and left without seeing the eruption.

It took two days to return to Kilauea. Jaggar was still suffering physically from the cold. A doctor was called. He prescribed rest and medication. Jaggar spent the next week in bed.

The eruption of Mauna Loa continued. After a week of rest, Jaggar organized a second attempt to climb to the summit. This time he and Lancaster ascended from the southeast side of the mountain, using only mules as riding and pack animals. The climb took two days. By noon of the second day, they were plodding through snow. An hour later, they were standing at the top of Mauna Loa.

Mauna Loa has a summit caldera much like the one at Kilauea, except more spectacular. The walls are higher and steeper and there are only two ways down to the floor.

From their position, Jaggar and Lancaster could see a short line of lava fountains near the southern caldera wall. Occasionally, a jet of lava would shoot higher than the others. When that happened, Jaggar could follow individual fragments of molten red rock as they fell, passing through successive shades of red from cherry to claret and ending as black.

The men found a way to the base of the fountains where they collected a few of the falling fragments. Unlike at Kilauea where much of the material tossed into the air remains whole when it hits the ground, here in the rarefied air of Mauna Loa the ejecta floats away slowly like crumpled sheets of burnt paper.

After returning to Kilauea, Jaggar was obviously pleased with his success. Always thorough, after writing a brief account of what he had seen, he wrote a private note to himself. It was a reminder of how to be better prepared for the next eruption of Mauna Loa. The list included: never rush to the summit, always make sure everyone has a change of clothes and "use mules if possible."

———•———

The new year of 1915 began a few weeks later. It would be the first full year of war in Europe. And that brought an unexpected consequence to the Hawaiian Islands.

The price of sugar doubled during the first six months of the First World War. It doubled again during the second six months. The result was to make the owners of sugar plantations, such as Lorrin Thurston, wealthy. For an obvious reason, the war also deterred people in the United States from traveling to Europe. Instead, those who had a desire to travel turned their attention to the Pacific, greatly boosting the number of tourists coming to the Hawaiian Islands.

In February 1915 the SS *Great Northern*, built just the previous year, made its first voyage to the Islands, sailing direct from Los

Angeles to Hilo. Its sleek design and powerful engines cut the usual travel time from six down to an amazingly short four days. On that maiden voyage, the *Great Northern* carried 495 passengers. When it arrived in Hilo, almost everyone went to see the volcano.

In May the *Great Northern* brought a full complement of passengers, which included a special group, to the islands: 124 Congressmen and Senators and their wives and other relatives and friends. The official purpose was to determine what new legislation was needed to govern the Territory of Hawaii. This time the *Great Northern* sailed direct to Honolulu where the group met Thurston, who would guide them around the islands. Their first night was at the Moana Hotel in Waikiki. That was followed by a few days of touring sugar plantations and pineapple packing plants. Then to the island of Kauai where Thurston showed them the need for new harbor facilities at the main port of Lihue. From there, the group went to Maui where there were more discussions about sugar production and where a large luau was held. Then Thurston took the group to the island of Hawaii where they spent their last afternoon in the islands at the edge of Halema'uma'u.

Cooks from the Volcano House prepared a meal over volcanic heat. The guests sat at tables covered with white linen and they ate off chinaware and used silver utensils. As the luncheon progressed, both Thurston and Jaggar circulated among the diners, telling them of the need to make Kilauea—including the crater of Halema'uma'u—into a national park.

House Congressional hearings into the matter were held in Washington, D.C., the next February. The Honorable Scott Ferris, a Democrat from Oklahoma, presided. Jaggar had been invited and would give most of the testimony.

The former Harvard professor spoke for three hours, giving, as one newspaper reporter described it, "a stunning display of showmanship and scholarship." Jaggar began by drawing a comparison between the wonders of Yellowstone and the lava lake at Kilauea. "It is a lake of fire hundreds of feet long, splashing on its banks with a noise like the waves of the sea, while great fountains boil through it fifty feet high, sending quantities of glowing spray over the shore of hissing gases and blue flames playing through crevices."

He continued. "It may seem to you that in this account I am too superlative, but those who have seen this liquid fire in the crater pronounce it the most marvelous thing that they have ever seen." Jaggar then presented the committee with an enlarged photograph of the lake that showed streams of lava outlined by black rocky crags.

Now Congressman Ferris chimed in. "I saw that. I was there. I stood right there!"

Next, armed with more photographs and with charts and diagrams that showed the recent changes in the lake, Jaggar spoke of his years of living atop the world's most active volcano. He told them of the recent eruption of Mauna Loa and how he had fought through a blizzard trying to reach the eruption site and how lava flows from future eruptions will threaten Hilo, the second largest port in the islands. He told them of his recent trip to Japan and the eruption of Sakurajima. And he reminded them of the tragedy of St. Pierre. Making Kilauea into a national park would not only preserve a national treasure, he told the committee, it would also be a boom to tourism and, hence, to the economy of the islands.

Four days after Jaggar's testimony, the House Committee issued a favorable report on a bill to make Kilauea a national park. Two months later, the bill passed both the House and the Senate. On August 1, 1916, President Woodrow Wilson signaled his agreement by signing the bill, creating Hawaii National Park.*

As a corollary to Kilauea becoming a national park, on February 3, 1916, the day that Jaggar testified in front of the Congressional committee, the lake level was 420 feet below the crater rim. Wood, who made the trip that day to inspect the crater, recorded that the lake was "vigorously boiling." By early April, when the

---

* In 1961 the summit of Haleakala volcano on Maui was made into a national park. At the same time, the name of the park on Hawaii was changed to "Hawaii Volcanoes National Park," the name that is used today.

House and the Senate voted to establish the national park, the distance down to the lake was only 300 feet. On June 4, the lake level was 265 feet, the highest in four years, that is, since the establishment of the Hawaiian Volcano Observatory. Jaggar was at the crater that evening and recorded that the lake showed "increased fountaining and rapid streaming."

The next day, June 5, Thurston, the old revolutionary, announced that he and George Lycurgus, who had been a royalist, had agreed on a selling price for the Volcano House and that Thurston would be taking control of the hotel later that day. At 8:40 that morning, according to Jaggar's notes—he was standing on the crater rim at the time—the lava lake started to subside slowly. After two hours, it was down forty feet. The crater walls now started to fall inward, sometimes as great blocks, causing great clouds of cream-colored dust to rise from the crater. It was, according to Jaggar, "the most gorgeous display of volcanic clouds the write has ever seen." By nightfall, the crater floor had dropped an additional hundred feet. By the end of the next day, it was seven hundred feet, the lowest since 1894.

Thurston withdrew his offer and never tried to buy the Volcano House again.

————•————

As the floor of Halema'uma'u was dropping in June 1916, discussions were already underway, mostly in private, as to the future of the volcano observatory and whether Jaggar should continue as the director.

In a letter to Thurston, MIT President Maclaurin expressed a concern about his institution's continued involvement. In particular, Maclaurin wrote, "various rumors have reached here of [Jaggar's] domestic infelicities and allied troubles." In the end, Maclaurin informed Thurston that MIT would end its involvement at the end of its five-year commitment, that is, on June 30, 1917.

There was also a growing discontent between Jaggar and his assistant, Harry Oscar Wood. As early as April 1915, Wood wrote to a friend, saying, "Things are not going well here, either for me personally or scientifically, or for the present or future of this Observatory." He reported that he had "grown very tired and stale here" and that he felt covered with "a deep coat of rust, scientifically." Furthermore, in his opinion, "personal troubles have hampered and threaten to destroy Jaggar's usefulness, in his scientific capacity as well as in his promoter-administrative capacity here."

Arthur Day of the Carnegie Institution of Washington, who had collected gas samples with Shepherd and Lancaster at Kilauea in 1912, agreed with Wood. In a letter addressed to Wood he wrote: "It is in no sense surprising that you should have tired of the top of Kilauea as a continuous place of residence. The trouble, which is very real, is that few men can work continuously with either pleasure or profit in isolation." It was for that reason that Day had refused regular employees of the Carnegie Institution from involving themselves in "continuous volcano observation."

And there was Thurston whose concern was how to finance the work of the observatory beyond the original five years. He had raised the money to begin the observatory and had kept the observatory running, even contributing his own money when the observatory's bank account ran low. He had honored his five-year commitment, and the five years would soon be over. After his recent failure to buy the Volcano House, his interest in Kilauea as a commercial endeavor waned. If the observatory was to continue, a new source of money had to be found. Thurston found it in the form of Herbert Gregory, a professor of geology at Yale University.

Gregory made his first trip to the Hawaiian Islands and visited Kilauea in May 1916. He was in the islands because he had been chosen to be the next director of Bishop Museum, a position he would begin the next year. His first priority as the new director, which he announced publicly during his initial visit, was to take control of the volcano observatory at Kilauea.

Later, in a private meeting, Thurston and Gregory discussed the future of the volcano observatory. "What should become of Jaggar?" Thurston asked.

Gregory answered. "He would be a good man for explorations in Alaska."

———•———

As the fall of 1916 approached, it was the low point of Jaggar's life. He knew of Gregory's plan to oust him as director. The floor of Halema'uma'u had just collapsed. And such collapses were often followed by months—or years—of inactivity. But this one was different.

Two months after the June 1916 collapse a small lake of lava formed 450 feet below the rim. A month later it stood at 330 feet. On November 1, there were nine vigorous fountains creating a tremendous outpouring of lava that covered the entire crater floor. Now the lake level had risen to within two hundred feet of the crater rim.

On December 20, there came a surprise. Walter Spalding, a 1910 graduate of MIT, stopped at the observatory to see his former professor. He told Jaggar that he had just been to Halema'uma'u and that, without the use of ropes or a ladder, had found a way down and stood next to the lava lake, which was now barely a hundred feet down. Jaggar had waited four years for such an opportunity. It had come just six months before he was to end his five-year leave-of-absence and return to MIT. He had time to begin a new series of experiments.

There was another reason the next few months would be momentous for him at Kilauea.

He had met a woman.

CHAPTER ELEVEN

THE SCHOOL TEACHER

C harles Eagan was in San Francisco and he needed a lawyer, that is, he needed a lawyer who was willing to return with him to the Hawaiian Islands.

After the revolution of 1893, the forested land along the road that led from Hilo to the Volcano House was opened to settlers, the post-revolutionary government encouraging them to clear the land and grow coffee. By 1896 more than six thousand acres were under cultivation, most of them small, individual plots. The largest one, three hundred acres, was owned by Eagan. The first harvest came in the spring of 1898. It was a dismal one.

Wet weather caused most of the coffee beans to rot, and a boring beetle ravaged those beans that did mature. And so coffee growers were selling their lands at greatly reduced prices. And Eagan was anxious to buy those lands and consolidate them into a single large sugar plantation, knowing that the growing of sugar was still the quickest way to get rich in the Hawaiian Islands. But there was a problem. The politics as well as the legal system in the islands was under the control of a few people, mostly the descendants of missionaries. Eagan was a newcomer, having arrived in 1895. And that was why he needed his own lawyer to help him expand his holdings.

Eagan found him in the form of Guy Maydwell, a young attorney from Sacramento who had recently moved to San Francisco. Maydwell was having a difficult time of it. He had his own law office on Sutter Street, but he had few clients. In addition to his law practice, he worked at night as a bookkeeper in an accounting firm. He was open to new opportunities. And Eagan laid out a grand one.

It was June 1898 when they met. Four months earlier, the battleship *Maine* had exploded in Havana harbor, sending the United States into war with Spain. Two months later Commodore George Dewey defeated the Spanish Navy in Manila Bay, taking possession of the Philippines. As Eagan pointed out to Maydwell, the United States was in the process of annexing the Hawaiian Islands to provide a way station between the west coast and the Philippines for its Navy. And once the Hawaiian Islands were annexed, those who were already living and working in the islands would be the ones who would benefit most from the economic boom that would follow annexation.

Furthermore, as Eagan pressed the point, Maydwell would be assured of legal work. He would be Eagan's personal lawyer, preparing and filing the legal papers that would be required to acquire the land to start the new sugar plantation, a plantation that would be located near Hilo Bay along the road to the volcano because that land had already been cleared.

Maydwell, it is safe to say, was easy to convince. He contacted his sister, Ollie, who was also working in San Francisco, and she agreed to go with him. He also sent word to Isabel Peyran who he had known for several years and who was Ollie's best friend. Maydwell told her of the opportunity in the Hawaiian Islands. He also asked her to marry him. She said yes.

Peyran was the daughter of French immigrants who had come to California during the Gold Rush. She was born in a mining camp on the west side of the Sierra Nevada on August 17, 1874. The year after her birth the Peyran family moved to Sacramento where her father found work as a lumberman and her mother as a seamstress. When Isabel was six years old, her mother, Marie Peyran, citing "a failure to provide," was granted a divorce from Phillipe Peyran. The family situation grew worse a few years later when her father was murdered by thieves who broke into a house where he was sleeping.

Isabel attended school up to age fourteen. After that, she began her working career as a teacher. Her first assignment was at Lincoln Primary School in Sacramento where there was another first-year teacher, Ollie Maydwell. Ollie introduced Isabel to her brother. From that, a courtship developed, though exactly when is unknown.

As soon as he asked her to marry him, Isabel left for San Francisco. They married on July 2, 1898. Five days later, President William McKinley signed the joint Congressional resolution that annexed the Hawaiian Islands. Five days after that, on July 12, the Maydwells sailed for the islands.

They arrived in Hilo two weeks later. Eagan met them and moved them to what remained of his coffee plantation along the road to the volcano. Later that month, he and Guy Maydwell were in Honolulu where they watched the annexation ceremony and where Maydwell applied for a license to practice law in the islands. The license was granted in December. By then, an unknown buyer had purchased most of the former coffee lands. For months, there was speculation as to who it might be. Finally, in April 1899, he revealed himself. It was Lorrin Thurston.

Eagan was the sole holdout, refusing to sell his land to Thurston. How Thurston eventually got control of the land is a complicated story. Let it suffice to say that a year of legal suits and countersuits followed. In the end, the matter was settled out of the courts after Eagan purchased a racehorse from one of Thurston's business partners. Eagan raced the horse in Hilo on July 4, 1900, in a free-for-all race. His horse beat the competition easily. Eagan collected the prize money of $200 and the side bets. Then, after signing ownership of his land to Thurston, he sailed away and never returned to the islands.

Guy Maydwell was now without a client. What would he and Isabel do?

———•———

The Maydwells moved to Hilo where Isabel and Ollie opened a free kindergarten, the first preschool in Hilo, and Guy opened a private law office. He did divorces and researched land deeds. But whenever he went to court and Lorrin Thurston was the opposing attorney, Maydwell lost the case.

Fortunately for Guy Maydwell, the court clerk on the west side of the island in Kailua died and a judge hired Maydwell to be the new clerk. He left Hilo and moved to the other side of the island. Isabel stayed, taking charge of a boarding house.

As one of two attorneys on the west side of the island, Maydwell soon had most of the legal business offered by local ranchers and shipping agents. He again opened an office. He also entered politics, elected in 1902 as one of the island's two prosecuting attorneys. He acquired more legal clients. Isabel moved and joined him. They rented a large house in the community of Holualoa, high on the steep slope that looks over Kailua. The house was across the road from the local school where Isabel was hired to teach. They were now at the top of the local social strata. They attended formal and informal dinners. They enjoyed weekend picnics and were invited to taffy pulls. After years of struggling, Guy and Isabel Maydwell had achieved financial stability. But that ended quickly.

Six people taught at the school in Holualoa, including the principal, Nettie Scott, whose husband had built the schoolhouse. In August 1909, as the vice-principal, as well as one of the teachers, Isabel Maydwell filed charges against Scott with the school board, claiming the principal had acted with "cruel and inhuman treatment of pupils" and had "arbitrarily and arrogantly conducted herself toward her subordinate teachers." The school board met to consider the matter. Parents and students attended in support of Maydwell, many of them willing to speak on her behalf. But the board members refused to hear testimony. Instead, the members had already made a decision. They charged Maydwell with being insubordinate in filing the charges. And so her position at the school was terminated. That ended her teaching career. She never taught again.

A month after the incident with the school board, Guy Maydwell complained of headaches. He went to Honolulu to be examined by a local physician, James Judd, who diagnosed a brain tumor. Dr. Judd advised him to go to San Francisco for surgery, but Maydwell refused. Instead, on June 24, 1910, Judd performed the operation, removing a tumor about the size of a small orange that was wrapped around an optic nerve. For days, Maydwell hovered between life and death. Eventually, he grew strong enough to leave the hospital, but the operation had left him partially paralyzed. And so a second one was performed. His condition worsened. On September 10, 1901, Guy Maydwell died at Queen's Hospital in Honolulu.

His wife was now alone, widowed at age thirty-five. She returned to the island of Hawaii where she mourned his death for six months. Afterwards, she began a series of quick trips to Honolulu. On one trip she stayed several months, working at a post office, later as a bookkeeper at a small sugar plantation. But she always returned to Hawaii.

In August 1911 she went to see the volcano, staying three nights at the Volcano House. She took the trip to Halema'uma'u and saw the active lava lake. Frank Perret was then living in a small wooden hut at the edge of the crater. She might have met him—almost everyone did. She did sign a guest book that Thurston had left at the hut. That night, when she returned to the hotel, three quick earthquakes rocked the hotel, the shaking strong enough to rattle the windows.

She made her next trip to Kilauea in May 1913 and a third one in October 1914. It was during the latter trip that she first met Thomas

Jaggar. He had just returned from the eruption of Sakurajima in Japan and probably told her of his recent adventures. She returned to the volcano in February and again in May. By then, their relationship was more than just causal: On his monthly bill, in addition to paying his normal board and room and two dollars for laundry and a dollar for a box of cigars, he also paid three dollars for "half a room for Mrs. Maydwell." She had stayed at the hotel for three nights, traveling with a female friend.

Then, after the year 1916 ended and 1917 had begun, there came two catalysts in their romance. One was an eruption of Kilauea. The other was none other than Lorrin Thurston.

————•————

After Spalding had reported to Jaggar that the lava lake could be reached, Jaggar made plans for new experiments to measure lava temperature. In preparation for those experiments, on January 3, 1917, he made his first descent into Halema'uma'u, writing of the day, "I was first privileged to walk over the hot lava and stand on the ramparts a few feet from the surging lake."

He was accompanied by William Twigg-Smith, Thurston's son-in-law and the president of the Hawaii Tourist Bureau. The two men made their descent down the east side of the crater, Jaggar carrying a sledgehammer, which he had named "Excelsior," a reference to a popular poem by Henry Wadsworth Longfellow and which meant "onward and upward," knocking away overhanging rocks as he went.

Standing on the crater floor, they made their way carefully, testing with their boots the footing of every step. They found a place where stiff toes of lava were pushing out from the base of a larger molten mass. Here Jaggar used a long wooden stick he had brought with him to poke at the toes, discovering they were covered with a glassy membrane "stiff as leather and flexible."

The two men resumed walking, coming to the base of a long low hill that was smoking furiously. They scrambled up the side, a rise of about twenty feet, finding the surface exceedingly slippery and covered with black, glistening shards of fresh glassy lava, the shards cutting and scarring their boots.

Reaching the top, they saw that they were standing on the crest of a low ridge that was acting as a dam that was holding back the lake of molten lava, the dam keeping the molten material from flooding lower parts of the crater floor. Red churning lava was just a few feet below them. In one direction the ridge ran up against the crater wall. In the other direction it ended at the edge of a large peninsula of solidified rock that extended out into the lake. Jaggar decided it was there at the end of the peninsula that he would measure directly the temperature of the lava lake.

The idea was a simple one. He would fill one end of a long iron pipe with Seger cones. Seger cones had been used by the porcelain industry for many years to measure the high temperature of furnaces. Each cone had a mixture of clay and limestone and other minerals and was formed into the shape of a small pyramid. The precise mixture determined the temperature when the pyramid would lose its shape and sag, or fuse. By putting Seger cones of different fusible temperatures into an iron pipe, then plunging the pipe into the lava, retrieving it and examining which cones had fused, Jaggar could determine the temperature of the lava lake.

Two days were required to haul by truck bundles of iron pipe to the edge of Halema'uma'u. Another day was needed to lower the bundles by rope to the crater floor, Jaggar assisted by Twigg-Smith and Lancaster. Thurston, who was financing the experiment, arrived on the last day and also assisted. The next day, January 7, would be a Sunday. Thurston suggested that, instead of beginning the experiment the next day, they rest and bring their families and friends and lead them down into the crater.

When tracking the progress of a romance, it is important to realize not only what a couple does, but also what they try to hide. In the case of the romance between Isabel Maydwell and Thomas Jaggar, it is what they did *not* record in a personal diary and in field notes on January 7, 1917, that is important.

In his notes, Jaggar listed those who were members of "the large party" that descended into Halema'uma'u. There was Thurston and his wife and daughter. There was also Thurston's business partner, Walter Dillingham, and his wife and son. A Dr. Adams was included. One presumes he was a guest at the Volcano House. And there is Alex Lancaster. At the end of the list, Jaggar wrote "others," as if to indicate that the others lacked the importance of the other members. Then, sometime later, in the narrow space between lines, he inserted "Mrs. Maydwell." By itself, the addition would mean little if she had not done something odd.

In a notebook that she used as a diary, she removed the first few dozen pages by cutting through them with a razor blade as if she was eliminating her past. Then, on what was now a new first page, she wrote an entry for January 7, 1917. She records the names of every person who was on the outing except one, Jaggar, even though he was leading the way and explaining the recent activity of the lake and taking photographs. Admittedly, the omission is a minor act.

And, yet, one feels it was deliberate, perhaps, because this day was an important point in their romance, producing a personal feeling that couples try to conceal. Afterwards, she no longer went on extended trips to Oahu. She lived most of the time at the Volcano House, working as Jaggar's unpaid assistant, typing notes and learning to read seismic records.

And he, though immersed in his work at the lava lake, was preparing to do what was necessary to remain at Kilauea.

———•———

Three days of rain followed the Sunday outing. Finally, on Thursday, January 11, the weather cleared. At 9 A.M. Jaggar and four other men stood on the rim of Halema'uma'u and were ready to work. The other four were Lancaster, Thurston, Twigg-Smith and Joe Moniz, a local man.

Before starting the descent, Jaggar studied the crater floor through binoculars. He could see the long ridge where he and Twigg-Smith had stood the previous weeks and the peninsula that extended out into the lake. Since that trip, much of the peninsula had been spattered by fresh lava. No matter. With a matter-of-fact tone in his voice, Jaggar called to the others. "Come along and let's get to work." He led the way down.

As soon as the five men reached the crater floor, Jaggar had them carry the bundles of iron pipe to a place at the base of the long ridge and near the peninsula, a place he thought was relatively safe from being run over by molten lava anytime soon. The ends of the pipe had already started to corrode. And so Jaggar had the others rub the pipe ends with steel wool while he and Thurston climbed the ridge and walked out onto the peninsula.

The entire area was steaming more than it had the previous week. The ground was also noticeably hotter, enough to burn the leather soles of their boots if they were careless and stood still. The only solution, as Jaggar and Thurston soon discovered, was to keep raising one foot, then the other to avoid the smell of burning leather and to give the steel nails on the soles of their boots a brief time to cool.

Thurston had brought a small log. Standing atop a small promontory along the edge of the peninsula, he tossed it onto the lake. A few small flames sputtered out from beneath the log where it set atop the lava. There was too little oxygen in the air directly over the lake for the log to flare into flames. Slowly the motion of the lake surface carried the log away until it got caught in a down-going lava stream. Eventually, it was pulled under, a puff of white steam marking where it had disappeared, followed by a small lava fountain that continued for several minutes.

Back with the others, Jaggar had the men screw together several lengths of pipe. At one end he attached an empty iron capsule that would hold the Seger cones. He placed six cones inside with fusible temperatures that ranged from 1,810 to 2,100°F. He then closed the capsule with an iron screw cap. To the other end of the pipe he attached a rope that would be used to pull the pipe out of the lava lake.

When all was ready, Jaggar led the way, the other men following and carrying the assembled pipe, climbing onto the ridge, then out onto the peninsula. The men could hear their boots crunch through thin crusts of new spatter.

Jaggar positioned himself at the end of the peninsula farthest from the ridge, standing within four feet of the edge of the lava lake. The heat radiating off the silvery satin surface of the lake was incredible. Breathing was difficult. Fortunately, he did not smell any noxious gases. Shouting, he directed the others to push the end of the pipe with the iron capsule to him. He grabbed onto it, steadying himself as best he could, then plunged the end of the pipe into the lava. It was instantly swept aside by a strong undercurrent, nearly pulling him off his perch.

He crouched down, holding the pipe in place with one hand, which was covered with a heavy leather gauntlet glove, while shielding his face with the other, shifting his feet constantly. He dropped his free hand only once, briefly, to look across the fiery lake. The air was surprisingly clear. Lava fountains seemed to be in every direction. Surprisingly, there were no sounds, the heated air somehow muffling the noise.

Yellow flames tinged with pale blue danced frantically across the surface of the lake. The hot air shimmered so fast that the huge distant black crags of solidified lava seemed to oscillate. He decided not to take any deep breaths lest the nearby searing furnace-like air singe his lungs.

The four other men stood far away, holding onto the rope tied to the free end of the pipe. They waited for a signal. "How long?" Jaggar shouted. "Six minutes," someone called back. Jaggar sprang to his feet and told the men to pull with all they had.

The four men wrapped their gloved hands around the rope and leaned back and pulled. But lava hung tightly to the submerged end of the pipe and it would not move.

The men began to pull in rhythm. Slowly, imperceptibly at first, the pipe started to slide. At last, it came up in a rush.

The pipe was carried to a safe place. The capsule had been submerged to a depth of only a few feet. It was covered by several inches of still-hot lava. The men cooled the lava with water from their canteens. Then Jaggar chipped away at the crusted lava with an ice axe. As soon as it was clean, he unscrewed the cap and pulled out the six Seger cones. None had fused. The capsule had not been immersed long enough for the cones to feel the full heat of the molten lava.

Over the next months Jaggar tried the experiment several times again. On one occasion he enlisted the help of five men of the United States Army infantry who were visiting the volcano. One of the soldiers, who had been in the Spanish-American War, remarked that standing near the lava lake made him more nervous than any gun battle he had been in.

Jaggar tried different ways to preheat the capsule before immersing it. The most promising method was to hold it inside one of the red-glowing caves found around the edge of the lava lake for ten minutes, then immerse the capsule beneath the molten lava for five more. In one such experiment, all of the cones fused, except one. From that single experiment, Jaggar concluded the temperature of the molten lava lake at Kilauea was 2,010°F (1,100°C), almost 100°F hotter than the temperature measured six years earlier by Perret and Shepherd. Countless measurements of lava temperature, using a variety of techniques, made since then have confirmed Jaggar's result.

Standing next to the lava lake for hours at a time, he naturally wondered how deep the lake was. Those who had seen the lake and drawn cross-sections—including James Dwight Dana of Yale University and Ernest Shepherd of the Carnegie Institution of Washington, as well as many others—had assumed the liquid lava extended deep into the volcano. But might the lake have a bottom, one that could be measured? From his years of observing the lake, he thought so. And so he devised a way to measure it.

The first attempt was made on January 23. Nine volunteers, all guests from the Volcano House, assembled at the lake's edge. Jaggar had them screw together ten lengths of twenty-foot-long half-inch-diameter stainless steel pipes to make a single long one. He then had the nine men lift the long pipe and walk forward with it until each man stood within a few feet of the lake's edge. Some of the men became nauseous from the suffocating heat. Others would later recount how the skin on their faces blistered immediately from the radiant heat of the lake. But all nine stood firm and waited for instructions. The instructions came quickly.

Jaggar stood at one end of the long pipe, holding it steady, then plunging it into the lake of liquid lava. He chose that part of the lake because it had a current. As soon as the pipe touched the lake it began to slide away in the direction of the current and the nine volunteers. As the submerged part of the pipe passed each man, Jaggar came running by, telling the man to drop his section of the pipe and stand back.

Down went the pipe until it could go no farther. Grabbing hold of it, Jaggar could feel that the submerged part had encountered resistance. He tried to push the pipe in deeper, but that caused it to arch up. His hand shot up, signaling the others that it could penetrate no farther. Sixty feet of the pipe was submerged. Now began the difficult work of retrieving it.

Hand over hand the nine men pulled, each one grabbing the hot steel, enduring the pain as hundreds of tiny shards of volcanic glass cut through their gloves and sliced into their hands. Lava had congealed around the immersed section of pipe, making it difficult to pull out the pipe. The men could pull out only the first twenty feet; forty more feet of steel pipe was still immersed when Jaggar told them to stop. A block and tackle was rigged and the men again tried to retrieve the pipe, but it was hopeless. And so Jaggar had the men unscrew the sections that could be reached. The last two lengths were abandoned.

Someone had the idea to place his hand over the open end of the stuck hollow steel pipe. He felt cold air coming out in regular pulses. He placed his ear near the open end and told the others that he could hear a sound similar to the rhythmic beating of an exhaust pipe when a car motor is running. Everyone took a turn, surprised by both the cold air and the fact that there was anything to hear.

Jaggar made a second attempt to sound the lake the same day, this time from a place a few hundred feet from the original spot.

Another long steel pipe was constructed and it, too, was slid into the lava, then it stopped. Again the men pulled out the pipe, this time, retrieving the entire length. And, again the pipe had gone down about sixty feet.

Jaggar repeated the experiment four more times on four different days at four different places around the edge of the lava lake, each time with a different team of volunteers. And each time the depth was the same.

From these crude experiments and from his years of closely watching the circulation pattern of the lake and making daily measurements of the lake's changing level, Jaggar came to the following conclusion. He discarded the old concept that the lake was merely the top of a large cylinder of molten material that continued deep into the volcano. Instead, he had shown that the lake was a shallow feature. Most of Halema'uma'u was not filled with molten lava, but with a plug of pasty lava—in his words, "an impenetrable pudding"—laced with channels. It was through these channels that hot liquid lava was fed upward from a deep reservoir and supplied molten lava to form the lake. And it was through similar channels that the fluid lava circulated back down.

————•————

On January 12, the day after Jaggar made his first attempt to measure the temperature of the lava lake using Seger cones, MIT President Maclaurin wrote to him acknowledging the receipt of his resignation from MIT and to tell him that MIT had no future plans "for carrying out scientific work in Hawaii." Three days later, the board of trustees of Bishop Museum issued a report that supported Gregory's proposal for Yale University to take over the operation of the Hawaiian Volcano Observatory after MIT's support ended on June 30 and for Bishop Museum to fund the effort. To Jaggar, it must have seemed as if there was a conspiracy to oust him.

And there was. Thurston had already lent his support to Gregory's plan, writing to Jaggar, "I have great hope that the combination can be made between Yale University and the Bishop Museum." Jaggar had reacted magnanimously, writing back, "If necessary for the good of the work I would eliminate myself entirely provided I was assured that that ideal of systematic recording of the volcanoes would be preserved at the Observatory along the line which I started."

But, as the months passed, his public attitude changed. He now wrote to Thurston of "the seemingly high handed demands of Yale" and of Gregory's call for the "unconditional surrender of the Observatory." But how could he stop what seemed to be inevitable.

Fate intervened.

Some members of the Hawaiian Volcano Research Association, the group that officially financed the observatory, wanted Jaggar to continue, but Thurston was the dominant member and, as in most things in his life, he got his way. But in June 1917 he became ill and went to New York for medical treatment. That opened the door for other members of the Association to act.

They met while Thurston was away, having a quorum so that all votes were binding. To discard Gregory's proposal outright would have been too drastic, and so they sidestepped the issue. They voted to keep control of the observatory, even though there would soon be no money to operate it, including the paying of salaries. And they voted to keep Jaggar as director and to dismiss Wood, who Gregory was proposing as the man to replace Jaggar.

Later one of the members of the Association sent an unsigned note to Jaggar about the meeting and Wood's dismissal. "I cannot help feeling glad that our little friend is to go. I have known for a long time that his attitude towards you was one of never neglecting a chance to sneer at your work, and your private affairs." At the end of the note, the writer added, "Thurston will not like it, but he is not the whole cake."

When Thurston returned, the decision had been made. He continued as a member of the Association, though reduced his financial support of the observatory.

Jaggar, now assured that he would continue as director and remain at Kilauea—though his salary would no longer be paid—made an important decision. He and Isabel Maydwell decided to marry. He wrote to a friend about the coming nuptials.

"Truly happy? Yes, I am truly happy, and the harmony is love, labor and liberty. I am going to be married. . . . The lady is Mrs. Maydwell who has been helping me in doing all the volcano record books, records which have won out to keep the Observatory going."

The wedding took place on Monday, September 17, 1917, at a private home in Hilo. The couple exchanged vows twice that day, once in a public ceremony, then in private. The public ceremony was done in the presence of a few close friends. The private one was between them: They agreed not to involve each other in their respective families, or, as he later stated it, "We agreed not to embarrass each other with kinfolk."

The next afternoon Thomas Jaggar was again standing at the crater edge taking notes. Two spatter cones on the crater floor were "spurting high with puffing noise." The north arm of the lake had "cracking and floundering crust."

The next day Isabel began an apprenticeship.

Initially, in observatory reports, her husband identified her as "general assistant," "recorder" or "skilled mechanic." As the months and years passed, she became much more. She learned to adjust and repair the seismographs, to read and describe earthquake tracings— Was the source beneath Kilauea or Mauna Loa or was it a distant quake?—and to describe subtle changes taking place at the lava lake.

She began by taking notes, her husband dictating to her. For example, on October 20, about a month after their wedding, her husband dictated five pages of notes to her. Four days later, it was eight pages. Each page was later transcribed and typed by her. She started taking her own notes. This went on for nearly six weeks. On December 11, Isabel made her first solo trip to describe the crater.

It was night, about 8 P.M. when she arrived. The sky was clear and moonless. She noted that a slight wind was blowing from the north. She stood at the southeast edge of the crater, just outside a dense plume of smoke rising from the crater. She could see faint bluish flames all around the edge of the lake. She noted the red glare of the lake was particularly bright that night, indicating, so she speculated, that the glassy crust was probably thin. There were two streams of molten lava creeping across the crater floor. One was contained within the main lake and flowing to the southeast toward her. The other was a stream originating from a far cove where lava fountains were bursting. She ended her first report succinctly: "Heat from the lake was strong."

———•———

Soon after their marriage, the Jaggars built a house at Kilauea. They chose a site just below the top of the cliff near the observatory building, out of sight of the hotel and on the leeward side of a rock wall that would protect them from cold rain carried by trade winds over the volcano's summit. They named their house *Kualono*, which means "near the mountaintop."

"I think you would enjoy staying awhile at Kualono," Isabel once told an interviewer, "sitting as it does on one of the down-sunken ledges of the crater wall, among trees and ferns and native plants."

Their house was comprised of three separate buildings connected by wooden walkways, each building at a slightly different level.

To reach the house, one descended a steep, narrow, zigzag path. A railing had been built to help visitors down. At the bottom, one stood on a wooden platform between two large buildings. The building on the right was the main one and was entered through a large sliding glass door. Inside was a single large room furnished with furniture made of rattan. Bookcases stood along one wall. A wood burning stove provided heat, though it was seldom used: The Jaggars had built their house over a steam vent, and the wooden floor was usually warm to the touch.

Beyond the main building was a smaller building that housed a pantry, a kitchen and a small dining area. A kerosene stove with four burners was used for cooking.

Back at the wooden platform and off to the left, opposite from the main building, was the third building. This one had two bedrooms and a bathroom with an iron tub, an iron basin, a toilet and a small kerosene stove to heat water. The master bedroom was the room farthest from the other two buildings. During the day it served as a private office. At night, the Jaggars slept in a Murphy bed that was pulled down from the wall. From the bed they could look across the room at a wall that was covered from floor to ceiling with glass panes. And through those panes they could see all of Kilauea caldera, including Halema'uma'u, and the distant profile of Mauna Loa.

Every night during their first two years of marriage, their bedroom was filled with red light reflected off the cloud that hung over the lava lake. During those two years, the scene was interrupted only once.

It was two hours after midnight on November 28, 1919, and Isabel Jaggar, the lighter sleeper, was awakened by the shaking of an earthquake. She looked out the window across Kilauea and saw the red glare of the lava lake brighten suddenly, then go dark. The floor of Halema'uma'u had collapsed.

She then turned and looked at her husband and saw that the volcanologist was still asleep.

CHAPTER TWELVE

THE LAVA LAKE

C harles Marvin never saw Kilauea—in fact, there is no record that he ever visited the Hawaiian Islands—and, yet, in a lifelong desire to extend his personal influence and involve himself in as many different scientific fields as possible, he made an important decision that would lead to understanding how Kilauea—and other volcanoes—worked.

In 1891 Marvin was a "junior professor of meteorology" in the Signal Corps of the United States Army when that part of the military morphed into the United States Weather Bureau. His expertise was the development of instruments that could automatically record meteorological data, such as air temperature, humidity and barometric pressure. Marvin excelled at his work. In 1903 when the Bureau acquired its first seismograph, an early Omori-type instrument that relied on the swinging of a long horizontal pendulum, it was put in his care. The thinking was: Since the Bureau was already collecting meteorological data at hundreds of stations across the nation, it might as well also collect seismic data.

Marvin installed this first seismograph owned by the United States government in the basement of the Bureau's office in Washington, D.C. Years passed as he carefully adjusted and maintained it, the instrument recording only the occasional, barely perceptible wiggle on a sheet of paper attached to a slowly rotating drum. Then, on the morning of April 18, 1906, the pen that normally traced out a straight line on the paper suddenly began to swing. And it did so for nearly thirty minutes.

Later calculations based on this seismic record would show that the ground surface around Washington, D.C., that morning had vibrated back and forth nearly half-an-inch, though the rate of the vibrations had been too slow—each swing back-and-forth had lasted a few seconds—for anyone to notice. That is anyone except Marvin who had the seismic record to prove it. Moreover, as it was soon determined, the source of the vibrations had been a train of seismic waves that had originated a continent away and had been caused by an earthquake near San Francisco. The seismic waves there had been so strong that some of the city had been destroyed. It took more than an hour for news of the devastation to arrive in Washington, D.C., via telegraph. The seismic waves had made the passage across the continent in just seventeen minutes.

Intrigued by what he had seen on the seismograph, Marvin began immediately to lobby Congress to make seismological studies a part of the work of the Weather Bureau. It took years of such work to convince Congress. Not until 1914, a year after Marvin was appointed Chief of the Weather Bureau, did Congress approve his request. Less than a year later, by geologic coincidence, Mount Lassen, a volcano in northern California, exploded, setting forest fires and sending up a column of volcanic ash that was seen for hundreds of miles. That event caused Marvin to set his sights on expanding the work of the Weather Bureau to include volcanoes.

Marvin brought Jaggar twice from Hawaii to Washington, D.C., to testify in front of Congressional committees in support of expanding the Bureau's work. In May 1918 both the House and the Senate acceded to the request and voted to include the study of volcanoes in the Bureau's work, but with a provision: The country was then at war in Europe, and so, because war brings uncertainty, Congress mandated that no funds would be approved for volcano work until the war ended.

That came on November 11, 1918. Two months later Congress approved the first federal funds for volcano research. On February 15, again with the approval of Congress, the Hawaiian Volcano Observatory became part of the federal government. Jaggar continued as the director. And Marvin sent one of the Bureau's longtime employees to be his assistant.

Ruy Herbert Finch was a gangly fellow. He stood nearly six feet tall and weighed barely 160 pounds. He had joined the Weather Bureau in 1910 at age twenty, serving at six different stations during the next four years. Each transfer to a new station was at his request. "This man does not like to be tied down to one location," wrote one of his supervisors. In evaluating his work, the same supervisor wrote that Finch was "too hurried for accuracy" and "had a roving disposition." Neither characteristic was desired by an organization that prided itself on accuracy and where most of the work was routine.

Finch was admonished officially at least once for his carelessness, a note in his personnel file reading: "The [afternoon] telegraphic report of June 9, 1913, contains an error in connection with barometric observations, the THIRD of its kind made by you within the past six months." But Marvin, now Chief of the Weather Bureau, saw promise in Finch. In 1915 he transferred Finch to Washington, D.C., and gave him a new assignment: the writing of seismological reports.

The new work fascinated Finch, so much so, that the twenty-five-year-old enrolled in nearby George Washington University to study physics. He was one year from graduation when, in 1918, the United States entered the First World War.

Finch enlisted in the United States Navy and was sent immediately to Ireland to be a weather forecaster and to help work out a scheme so that Navy pilots could spot German submarines in all weather conditions. Finch flew several times as an observer, but he never spotted a submarine. Others in his group did and, subsequently, nine submarines were sunk.

As soon as the war ended, he was sent to New York to wait for his discharge. He wrote to Marvin, asking if he might contact Finch's commanding officer and tell him that Finch's services were much needed at the Weather Bureau. "It won't hurt if you have to contort the truth," Finch added.

Finch got his early discharge. And Marvin offered him a choice. He could continue his university studies and become a researcher and administrator in Washington, D.C., or, because of his previous work on seismological records, he could move to the Hawaiian Islands and assist Jaggar at Kilauea. Finch, a young man who was still restless, chose the volcano.

He arrived on August 3, 1919, certainly surprised by Jaggar's latest undertaking. Jaggar had stationed himself at the edge of the lava lake and was in the middle of what he would later say was "by far the most interesting experience I ever had with volcanoes."

It was well known that ocean tides rise and fall twice a day because of the gravitational attraction of the sun and the moon. The question that Jaggar was trying to answer was: Might the lava lake at Kilauea also have tidal oscillations?

He began his lava-tide measurements at noon on July 21. Then, for the next twenty-eight days, one complete lunar cycle, he measured the height of the lake every twenty minutes using a surveyor's transit to determine the vertical angle between the top of the lake and white marks he had painted high on the crater wall. Measurements were day and night, regardless of weather or volcanic conditions.

It was a herculean task. And Finch arrived right in the middle of it. The lava lake then consisted of three circular pools, each one more than a thousand feet across, the pools positioned like the leaves of a three-leaf clover. The lake was then at its highest stand in more than twenty years, just a few feet below the crater rim. Jaggar set up the transit near the crater's edge. He had a shelter constructed to protect the transit from direct sunlight and from rain. It consisted of a large square piece of canvas held up by four heavy timbers. Occasionally, because of the heat of the surrounding solidified lava, one of the timbers caught fire. When that happened, the flames were doused by someone grabbing one of the pots of boiling water that were scattered around and used for cooking.

Jaggar was assisted by Isabel and by two young men who were visiting from Boston. One of the young men had told his father that he wanted something exciting to do while he was in the Hawaiian Islands. The father suggested that his son contact Jaggar.

Finch joined the other four in making the measurements. Jaggar worked out a schedule so that each person took a turn at reading the transit, resting, working as the recorder, and resting again. A bed with an iron frame was brought from the Volcano House so that they would have a place to lie down. The iron frame became so hot that heavy blankets had to be thrown over it to keep the occupants from getting burned.

Guests from the Volcano House came down at all hours of the day or the night to see what was happening. One of those guests who was particularly talkative and who peppered Jaggar with questions asked if the work was dangerous. As Jaggar remembered it, before he could answer, "I heard the hiss of escaping gas behind me and the pounding noise of highly charged lava. At the same time a bright flow spread over the landscape." He and the guest turned just in time to see "a fountain of liquid fire spurting" from a crack a few hundred yards away. The guest, suddenly quiet, did not wait for an answer to his question. Instead, he departed immediately and, so Jaggar was later told, checked out of the hotel and left the island.

In all, more than 27,000 measurements of the level of the lava lake were made during the twenty-eight days. In his assessment of the measurements, Jaggar was certain that they "reveal the extraordinary truth that a systematic tide lifts the lava in Kilauea crater so that the liquid is high in the forenoon and low in the evening." Others who studied the same measurements were not so certain.

Ernest Brown of Haverford College in Pennsylvania, an authority on the motion of the moon and the calculation of tides, heard of Jaggar's accomplishment and wrote to him, asking if he could examine the measurements. Jaggar sent a copy of all 27,000 measurements to him. In his final assessment Brown praised the volcanologist for conducting work "with the greatest care and under difficulties which can readily be imagined." But Brown had come to a different conclusion. After the application of his considerable mathematical skill, Brown saw no evidence of a lunar tide and thought the vertical movement of the lake level by the moon's gravity was no more than "an inch or so."

In other words, if a lava tide did exist, it could not have been detected by using a surveyor's transit.

———•———

The first written description of the lava lake was by Reverend William Ellis, an English missionary, who visited Kilauea in August 1823. He traveled to the volcano with eight Hawaiians and three local missionaries. Of this first recorded trip, Ellis wrote: "After walking for some distance over a sunken plain . . . we came to the edge of a great crater, where a spectacle, sublime and even appalling, presented itself before us."

Covering the floor of the crater was "one vast flood of burning matter, in a state of terrific ebullition, rolling to and fro." Ellis watched in awe as an "agitated mass of liquid lava, like a flood of melted metal, raged with tumultuous whirl." He asked the Hawaiians how long this had been going on. They answered that the volcano had been active from the beginning of time, in their words, *mai ka po mai*, "from chaos till now."

The first hint as to how Kilauea worked was recorded by Reverend Artemus Bishop who visited Kilauea with Ellis in 1823 and returned and saw the lava again 1825. The scene had changed greatly. The caldera—the "sunken plain" mentioned by Ellis—was much shallower, filled, by Bishop's estimate, with more than four hundred feet of recently erupted lava. John Honoli'i, who had been raised on the island, was with Bishop and informed the Reverend that "after rising a little higher, the lava will discharge itself towards the sea through some aperture underground." Such a "discharge" of lava did occur years later.

Titus Coan, an American missionary who lived on the island for many years, was on Oahu when the event happened. When he returned he was told what he regarded were fanciful tales of "the fiery matter" at Kilauea raging "like an ocean when lashed into a fury by a tempest" and of the ground shaking "with maddening energy." He went immediately to the summit to discount such claims.

He was surprised by what he saw. He had been to the summit of Kilauea several times, but, now, as he wrote, "Not a particle remains as it was when I last visited." The lava lake was gone and in its place was a deep steaming crater. He was told that many miles to the east lava had gushed out and flowed into the sea. Again he went to investigate.

It took him days to make his way through a dense forest. Finally, he arrived at a spot where the crust of the earth had been broken by fissures. And from those fissures had erupted a "mighty smoldering mass" of fresh lava.

He traveled in the direction the lava had flowed, finding that it led to the sea, as Honoli'i had predicted. Here Coan imagined what the scene must have been just weeks earlier, the lava "leaping a precipice of forty or fifty feet, poured itself into one vast cataract of fire into the deep below, with loud detonations, fearful hissings, and a thousand unearthly and indescribable sounds."

Now, eighty years later, Jaggar was devoting himself to Kilauea, watching and recording every change. Though he had failed to detect a lava tide, the next events would be easy to record because of the scale of activity. And because the volcano was, quite literally, ready to explode.

———•———

Jaggar had two ways to record the underground movement of molten lava. One was with the Bosch-Omori seismograph that, of course, could record earthquake shaking. It could also indicate a slight rise or fall of the ground surface by recording the accompanying slight tilting of the surface. To recall, it did this because the horizontal arm worked like a free-swinging hinged door: If the foundation of a house settles, the direction a hinged door hangs will shift. Likewise, if the surface of the ground tilts, ever so slightly, the rest position of the horizontal arm of the Bosch-Omori seismograph shifts.

The other way Jaggar could know that molten lava was moving underground was to use the lava lake as a giant pressure gauge. If pressure increased inside Kilauea, say, by the inflow of molten lava, then the level of the lake would rise. Conversely, if the pressure decreased, the level would fall.

The lake level was low after November 28, 1919, the night when Isabel had seen the red glare of the lava lake go out, indicating a sudden drop in the level of the lake. That was followed by more than two years of a general rise in the lake level. By the end of March 1922, the lava level was 178 feet below the crater rim. At the end of April it was only 57 feet, and on May 12 it was 49 feet. The next day a crisis began as the level started to drop.

"The sinking was steady but majestic," wrote Thomas Jaggar of the beginning of the event. His wife was more animated in her description. "The whole thing slipped down bodily like a plug rapidly enough but gradually and steadily so that [the floor of the crater] did not break up as it does in some subsidences." The slow collapse continued for several days. "Then it began to go down faster," wrote Isabel, "and we were thrilled." The Bosch-Omori seismograph recorded hundreds of earthquakes. Thomas and Isabel Jaggar went to stand at the edge of Halema'uma'u, to watch the collapse, enjoying it, as Isabel wrote, "When we weren't jumping back from the edge on account of the shaking of the ground."

Molten lava was last seen at the bottom of Halema'uma'u on May 20. And the floor of the crater continued to drop.

Attention was now drawn to the Bosch-Omori seismograph, which, like a hinged door, was seeking a new position of equilibrium, meaning, the ground was literally tilting beneath it.

From today's perspective, it is easy to overlook the importance of recording this movement—an indication the entire summit of Kilauea was subsiding and that molten lava was moving out of a underground summit reservoir and into another part of the volcano. But this was the first time anyone had measured such a subsidence *as it was happening*. Yes, Omori had measured a net subsidence *after* two eruptions, at Usu in 1910 and at Sakurajima in 1914, but not as either eruption was occurring. And then Jaggar succeeded at the next challenge: to record the slow refilling of a summit reservoir with molten lava. In fact, Jaggar would measure this cycle several times, setting the foundation for how volcanoes worked—thanks to the cycling of molten lava through reservoirs beneath the earth's surface—and, thus, providing inspiration for a countless number of future science fair projects by grade-school children who demonstrate how volcanoes work by inflating and deflating a balloon inside a papier-mâché mountain.

The summit of Kilauea was still subsiding on May 28 when, at about 8 P.M., the slow rumbling of an earthquake was felt across the entire island. Two hours later the Jaggars had just gone to bed when, as Isabel remembered it, "some people who live a mile down the road came to tell us that a glow could be seen in the sky." It was not coming from the summit crater, but far to the east within a dense forest. The Jaggars dressed and roused Finch who was staying at the Volcano House. The trio headed for the hotel's garage to get the observatory's automobile—Marvin of the Weather Bureau had purchased one for them—but there was a delay: The fan belt was broken. It took an hour to mend it by the light of a lantern.

They drove as close as they could get to the eruption, then found a trail that they followed until it turned away from the direction of the red glow, and, finally, blazed their own trail that took them over cracked ground and through dense obstructing vegetation, each person carrying and navigating with a kerosene lantern. They were chilled by a cold drizzle, got lost several times and were not at all sure where they might emerge. Finally, by 3 A.M., the three stood on the edge of one of the many small craters on this side of Kilauea. As they looked down, they could see a long line of spouting fountains midway on the crater wall from which a dozen ribbons of lava poured down. They spent the night on the edge of the crater, the rain continuing to fall, watching a puddle of lava accumulate on the crater floor.

"It was a beautiful sight," Isabel later recorded, "but was not large in extent nor very violent in action." In fact, it was so mild "that volcanologist Tom said that he thought it would be of short duration." And it was; the eruption ended soon after sunrise.

When they returned to the observatory, there was more excitement. The floor of Halema'uma'u was still dropping and was now nearly a thousand feet down, the deepest yet recorded, hundreds of feet deeper than after the collapse in 1894. Now the sides of the crater were caving inward, causing huge cauliflower clouds of ash and steam to rise from the crater.

"The old rocks of the walls began to go," wrote Isabel, "falling in with a roar that could be heard two miles away." Without sleep, the Jaggars spent the next day at Halema'uma'u, eating their lunch next to the crater. While they were eating, a large cloud floated directly over them and sprinkled them with volcanic ash. As it did, familiar with the story that almost everyone heard when they first came to Kilauea, Isabel "tried to imagine how the [Hawaiians] felt when many of them were overtaken, and some of them destroyed in this same region of the fall of the identical ash, gravel and pumice on which we were sitting." She was referring to the local story of a huge explosion of Halema'uma'u that eyewitnesses described as "a wonderful fire pillar standing in the sky, and at the top of this pillar a flame of fire was flashing, a flashing as of lightning." She wondered if what she was seeing now might be a forewarning of a dangerous explosion to come.

The eruption in the forest ended after one day. And the collapse of Halema'uma'u and the broad subsidence of the summit of Kilauea stopped after another week. Then molten lava returned to the crater. And the summit started to rise, indicating a refilling of the buried reservoir. The next year, in August, there was a repeat of the May 1922 event: collapse of Halema'uma'u, subsidence of the summit and an eruption on the lower slope of Kilauea, far from the summit. Again clouds of ash and steam rose from the crater. And molten lava again refilled the crater.

—————•—————

Two weeks later, on September 1, 1923, an earthquake struck Japan, demolishing buildings and causing massive fires to sweep through and destroy most of Yokohama and Tokyo.

More than a hundred thousand people died and more than a million were homeless. The United States government sent Thomas Jaggar to report on the earthquake damage. Isabel went with him.

They spent five weeks in Japan. While in Tokyo, they lived in a tent provided by the United States Consulate and shared a mess prepared by the United States Marine Corps. Near the end of the trip, the mayor of Tokyo sent them on a small steamer to the nearby volcanic island of Oshima. There were rumors the island was about to explode. Jaggar quelled the rumors, deciding that there was no immediate concern of an eruption.

They returned to Kilauea on November 10 to find that the lava level was three hundred feet below the crater rim.

Their six-year marriage, preceded by a year of courtship, had been a period of extraordinary activity at Halema'uma'u. The crater had collapsed and refilled four times. And they knew the time intervals between collapses. Forty months had passed between June 1916 and November 1919, then thirty months to May 1922, and fifteen months to August 1923. The next collapse should occur in another year or so, probably not earlier than the fall of 1924.

They had long delayed a trip to the United States mainland to visit friends. They also wanted to go to New York and talk to editors about writing books about volcanoes. And Thomas was anxious for a return trip to the Caribbean.

And so they must have been confident that, if they left immediately, they could complete their travels and return to Kilauea by June 1924, long before the next collapse of Halema'uma'u might be expected.

———•———

On January 7, 1924, a month after the Jaggars' departure, Oliver Emerson, a recent graduate of the University of Hawaii who had been working at the observatory for several months, made the daily trip to the edge of Halema'uma'u to record the activity of the lava lake. On that day, as Emerson noted, the lake level was two hundred feet below the rim. The lake itself consisted of two main pools that had come to be known as West Loch and East Loch. Three craggy islands comprised of dark solidified lava stood where the two pools joined. A few hundred feet from the molten lava of East Loch was a small lava pond that occasionally overflowed and sent red-hot liquid lava pouring into East Loch. On January 7, four such cascades of lava were streaming from the pool into East Loch.

Two weeks later, again making the daily trip to Halema'uma'u, Emerson recorded the lake level to be 110 feet below the crater rim. All three islands were now submerged. West Loch and East Loch were now one vast lake. In fact, this was the largest lava lake ever recorded at Kilauea. Its longest dimension was 1,500 feet. Its areal extent covered more than forty-five acres. There were also, as

Emerson noted, "countless tumultuous fountains" of incandescent red rock "dancing" across the lake's surface.

A month later, on February 22, the floor of Halema'uma'u dropped and the molten lava drained away. Emerson and Finch expected a replay of the cyclic events of the last few years. Instead, nothing happened, at least, not immediately. No clouds of steam or dust rose from the crater. No eruption of lava occurred elsewhere on the volcano. And no molten lava reappeared to fill the crater.

Two months later the silence was broken.

On the evening of April 21, forty miles from the observatory, at the eastern end of the island near the village of Kapoho, Henry Lyman was inside his house when he felt the slight shaking of an earthquake. By itself, it raised no concern. But, within minutes, a second shaking occurred, then a third. They became so numerous that Lyman was unable to sleep, and so he stayed awake all night counting earthquakes. By morning, when the flurry stopped, he had counted eighty-eight.

That afternoon, the shaking resumed and was more severe than the previous night. In fact, the shaking was almost continuous. One local person who decided to count earthquakes recorded 238 distinct events between 5 P.M. and 9 P.M.—an average rate of almost one a minute.

The next day, April 23, Finch, Emerson and Boles arrived to survey the area. They found dozens of new ground ruptures, some hundreds of feet long, running, in parallel fashion, in a zone from Kapoho to the sea—a distance of about a mile. At the sea's edge was a new lagoon where the ground had sunk as much as twelve feet overnight—an action that confirmed why the area was known as "Kapoho," a Hawaiian word that meant "a depression" or "a lowering."

The same day Finch announced publicly that an eruption should be expected soon at Kapoho. Instead, the earthquake rate diminished quickly. And no eruption happened. The volcano was again silent.

This time the silence lasted only one week. On May 1 the floor of Halema'uma'u started collapsing again, causing a funnel to form at the bottom of the crater. Small rock avalanches rolled continuously down the sides of the crater. The collapse continued, so that by May 6, the walls of the crater started to fall inward, enlarging the crater. By May 8 the bottom of the crater was more than six hundred feet deep and a roar of rock avalanches could be heard for miles. Clouds of dust and steam billowed thousands of feet into the air. On May 10 Boles pronounced the activity "the most spectacular seen in the Hawaiian National Park for years" and encouraged people to visit the park and stay at the Volcano House. That night the hotel was full. That night the volcano also exploded.

It happened at ten o'clock. In view of what was to come, it was a small explosion, rattling the doors and the windows of the Volcano House.

At daybreak, Finch, Emerson, Boles and a party of "interested people" staying at the Volcano House—that is how a newspaper reporter described them—drove in cars almost to the edge of Halema'uma'u, as close as they could go. New rock fragments were scattered all around. The largest was a four-hundred-pound block that had been blown a few hundred feet from the crater's edge.

That night Finch stayed at the crater's edge. For a brief period early in the morning he felt a spasm of earthquakes. And nothing more. There were no more explosions.

The next day and the next night were quiet, only heavy dust clouds rising from the crater. The following afternoon, May 13, Boles decided it was safe enough to lead a crew of seven who were filming a promotional movie for the Volcano House to the crater's edge.

Finch was standing at the highest point at Kilauea, a bluff known as Uwekahuna, a mile north of Halema'uma'u. At 4 P.M. he saw a double explosion, first near the center of the crater and second near the east side. He watched as rocks arched up above a rising dust cloud, one fragment going nearly half a mile into the air. He could hear the noise of those same rocks when they hit the ground around the crater rim and could hear a roaring coming from inside the crater.

Boles and the film crew were at the edge of the crater. "Escape seemed impossible," he would recall, "but we made a run for it." Flying stones and falling ash filled the air. "So terrific was their downpour," Boles would say, that "stone piled up around us." And they continued to run.

Boles was slower than the others, and he fell behind. At one point, his foot broke through a thin lava crust and he tripped and fell, cutting and scraping his hands and arms on the sharp surfaces. He staggered up and saw just ahead a shallow crevice large enough to get into. He dove into it, covering his face with his hands and trying to protect his head with his arms. Boulders bounced all around him. He thought a direct hit would have finished him.

The barrage of rocks lasted several minutes. After it ended, the others came back and found him. Two of them had been injured, but Boles was by far the worst. Back at the Volcano House, their wounds were dressed, Boles still suffering from the shock of the event. Finch, who had seen the explosion from Uwekahuna, was surprised anyone had survived.

The next day a huge explosion occurred at 9:05 A.M. It was followed by the rise of a dust-filled cauliflower cloud for an hour. At the end of the hour, as the rising of the cloud lessened, Emerson hustled down to the crater, finding that a 100-foot-wide ledge had fallen into the crater. The next day, again after an explosion, both Finch and Emerson went to the crater. They found hot rocks that "were still sizzling."

Another day passed. On May 16, fifteen minutes after "a good explosion with heavy cloud," as recorded by Finch, who viewed the cloud from the relative safety of the observatory, he and Emerson left for the crater. They found no significant change, just two small cracks that crossed the road a few thousand feet from the crater rim, an indication that the ground was weakening.

Thurston arrived that night. The next day there was another major explosion. As before, within minutes after it ended, Finch and Emerson, this time joined by Thurston, hurried to the crater. So much steam and dust was still rising and swirling around the crater rim that they were unable to make any significant observations. But they were convinced that there was a pattern to the explosive activity.

It was now May 17. During the last five days, Finch had noted, immediately after an explosion, which could last for up to an hour, the volcano was quiet for several hours and the crater could be approached.

And so it was that the next day, a Sunday, after a small explosion in the morning, Finch, Emerson and Thurston headed for the crater. This time several local people joined them. The group drove down in two automobiles. They kept the motors running in case they had to make a hasty departure.

Most everyone immediately set off and walked toward the crater rim. Thurston stayed next to the automobiles. Finch walked a short distance away to a sandy spit and sat down on a boulder that had been ejected the previous day.

At 11:07 A.M. Finch felt a wave of increased air pressure that was so strong it hurt his head. He jumped up and told himself, "Here comes a terrible one."

And terrible is was. Within seconds rocks were shooting out of the crater. Finch started to take photographs, then stopped when he saw the size of the cloud of dust and steam coming toward him. He ducked down behind a low cliff. The cloud rolled over him. A barrage of boulders, some weighing as much as three hundred pounds, dropped all around him.

After several minutes the explosion was over, and Finch ran back to the cars. All but one person had returned. The missing man was Truman Taylor, a twenty-four-year-old bookkeeper from a local sugar plantation who had been living on the island for only four months.

Someone said that Taylor had borrowed a tripod to take photographs and was standing at the crater rim when the explosion occurred. A search party was sent out. He was found 1,500 feet from the crater. He had been running. His legs were crushed and he was severely burned by hot ash. A raincoat was used as a stretcher and he was carried to the automobile. There was a second explosion, smaller than the first, and Taylor and the rescue party were showered by another fusillade of rocks. Fortunately, no one was struck.

They returned to the Volcano House where Taylor received first aid. He was then placed in a truck and rushed to a hospital in Hilo. One leg was amputated and it was planned to remove the other the next morning, but the loss of blood and shock was too much and he passed away during the night.

The same afternoon Boles declared the national park closed and ordered the closing of the Volcano House. The only people who remained inside the park that night were Finch, Emerson, Boles and Thurston. They continued to make notes of the activity and to maintain the seismographic records.

At 7:13 P.M. came the most awesome explosion of the series. It began the same as the others, but the explosion cloud continued to grow upward in leaps.

The first burst sent rocks a thousand feet into the air and barely a thousand feet out from the rim. Within minutes the air blast of a second burst was felt and rocks with incandescent trails were arching as high as half mile, then a third, and a fourth burst. Each time, rocks rose higher and fell farther from the rim until they were hitting the ground almost to the observatory more than two miles away. And then the explosion stopped. As Finch would write, "It was a real relief to have the bombardment cease."

Rain fell heavy that night and mixed with the volcanic ash that seemed to saturate the air. The result was a heavy downpour of mud.

Over the next week the intensity and the frequency of explosions lessened. The last explosion to throw rocks above the crater rim happened on May 24. Jaggar arrived in Honolulu four days later.

Thomas and Isabel Jaggar returned to New York from a trip to the Caribbean on April 24, just as the earthquake activity in Kapoho was waning. Their plan was to remain in New York until mid-June. Finch kept them informed of explosive activity. On May 18, after the death of Taylor, they decided to return home immediately.

They arrived in Honolulu on May 28. Thurston was at the dock to greet them. They were driven in an admiral's car to Pearl Harbor and taken to Ford Island where Professor Jaggar boarded a Navy seaplane—an Aeromarine 40, an open-cockpit, two-seat flying boat used during the First World War—that was waiting to take him to Hilo. This would be Jaggar's first flight.

His and a companion plane flew over Diamond Head and headed across the Molokai Channel. Jaggar looked down from a height of a few thousand feet at the beautiful pattern of whitecaps formed by trade winds. A half hour later, he was surprised that the surface of the sea was getting closer. Soon they were only a few hundred feet above the ocean. He was even more surprised that the other plane was far ahead. Finally, he felt the bump of wave after wave on the bottom of the pontoons as the pilot brought the seaplane to a stop close to a reef near Molokai.

The water was about fifteen feet deep, and a coral reef visible below. The pilot assigned Jaggar the task of throwing out an anchor and making the line fast to a cleat, while the pilot climbed up to the engine. The engine had been losing compression and could not keep up the required speed. The other plane came down and circled above them to see that they were safe, then went off. Meanwhile, Jaggar watched the water for sharks.

When the mechanic got the engine going again, Jaggar pulled up the anchor and they took off. They were airborne again, but for a short time. The engine again gave out and they came down. The pilot told him these were the first forced landings that he had ever made. The word "landing" seemed to Jaggar inapplicable.

They clambered up on top of the upper wing to wait for rescue. The wind was strong, and the seaplane drifted downwind for five hours. Finally, a small white boat appeared, coming from the island of Molokai. At the same time smoke showed that rescue ships were also coming from Pearl Harbor and from Maui. The Molokai boat reached them first and towed them to a nearby port, the downed seaplane pitching and taking such a pounding from waves that Jaggar did not think it possible that the mahogany hull and pontoons could hold together. But they made the harbor and tied up to a buoy.

The Navy tug *Navaho* from Maui came up. Its captain put up his megaphone to announce that his instructions were to take Professor Jaggar to Hilo. It was now an overnight trip. Jaggar was assigned a canvas cot in a lower level. All night long waves broke over the bow. A foot of seawater sloshed back and forth under his cot.

They made Hilo the next afternoon. Though wet and seasick, Jaggar was greeted at the wharf by a motion picture cameraman and a troupe of hula dancers who presented him with leis. Instead of five hours, the journey had taken thirty. He then sped by car to the volcano.

When he left in December 1923, Halema'uma'u was a shallow circular pit filled with molten lava and 1,800 feet across at its widest point. Now it was as much as 3,600 feet in diameter and more than 1,300 feet deep, the bottom a funnel of converging taluses made of rock avalanches from the crater walls.

Soon after he returned, Jaggar was asked what this phase of activity at Kilauea meant. He answered, "I don't know a damn thing."

———•———

Molten lava did return to Halema'uma'u in July 1924 for eleven days. And for brief periods in 1929, 1930, 1931 and 1932. In 1934 there was a month-long eruption. But none of this compared with the prolonged activity of the last few decades. There would not be another prolonged eruption at Kilauea for eighteen years, causing Jaggar to turn his attention to Mauna Loa.

CHAPTER THIRTEEN

MAUNA LOA

After twenty years at Kilauea, Isabel Jaggar was asked what she thought most people would be surprised to know about someone who lived on an active volcano. She answered:

"Some pity us, I do not know whether mostly for being poor or mostly for being feeble-minded. Most of our money goes for instruments or for some scientific thing which is needed at the observatory, but that's all right. We have some exciting experiences, watching sudden spurts of lava fountains in Halemaumau; seeing a whole mass of towering crags slowly and evenly slip down its throat and disappear below the rim."

She continued. "But the most exciting of all have been the expeditions up the side of Mauna Loa to hunt the source of a sudden outbreak of lava on its flank."

At such times, Isabel was in charge of organizing and packing the food and a cooking outfit. She also collected saddles, bridles and blankets and, if necessary, oiled and repaired them. She got out sleeping bags and warm clothing. All the while, as she remembered it, her husband kept "the telephone hot ringing up the different ranchmen on the slope of Mauna Loa" asking them for the latest report about volcanic activity.

As soon as everything was ready, they drove several miles to the end of a road, then transferred themselves and their equipment onto the backs of horses or mules, which they rode as far as a trail would take them. Then it was travel on foot across barren lava. "Everything must be packed on our backs," Isabel once wrote to her mother, describing a climb of Mauna Loa, "food, water, wood (if we take any), blankets, vacuum tubes for collecting gases, and cameras, cameras and more cameras. The water is for drinking only; clean faces are not at all necessary when one views the glorious display of Mauna Loa fountains. This is the life we have lived."

And so it was. The active lava lake had been the magnet that had attracted them to Kilauea. But it was Mauna Loa that presented the real challenge.

———•———

Mauna Loa is the world's largest active volcano.* Its long arched profile spans almost a third of the horizon as seen from the summit of Kilauea. The slope is remarkably gentle and nearly uniform. There are no ravines. The landscape is subtly mottled, a hint as to the volcano's character.

From almost anywhere that one is able to see the volcano, the sides are covered by long dark streaks that reach down for many

---

* To clarify what is meant by "largest" volcano: Mauna Loa is the most massive volcano that still erupts. The undersea volcano Tamu Massif in the northwestern Pacific Ocean has the most mass of any volcano, though its summit is over 6,500 feet below sea level and it has not erupted for nearly 150 million years.

miles. Each one is a recent lava flow, the blackness of a fresh flow not yet lost to weathering. As one's gaze moves higher, the streaks seem to narrow and increase in number. If one could get a bird's-eye view, it quickly becomes clear why: The streaks originate from the crest of a long, broad ridge that runs from northeast to southwest and crosses over the summit. It is from this crest that most eruptions originate.

The summit of Mauna Loa rises nearly 14,000 feet above sea level. Its base sets on a sea floor that is 15,000 feet deep, which means the total pile of volcanic rock is nearly 30,000 feet—or more than five miles—thick. The part of Mauna Loa that rises above the sea covers more than 2,000 square miles, an area almost the size of the state of Delaware. The submarine part of Mauna Loa covers nearly five times that area. A single eruption can pour out more lava in a few weeks than all the eruptions of Vesuvius since the destruction of Pompeii. If one wants to choose a single word to describe Mauna Loa it would be "immense." And that is its appeal.

The first person to record an attempted ascent of Mauna Loa was John Ledyard, a corporal of marines under the leadership of British explorer Captain James Cook. On January 26, 1779, Ledyard set out from the west side of the island with four shipmates and with about twenty Hawaiians as guides and porters. After two days, running low on provisions, Ledyard realized he was less than half the distance to the summit and so he abandoned the attempt.

Twice, in February 1793 and in January 1794, Archibald Menzies, a surgeon and botanist on an expedition led by Captain George Vancouver, another British explorer, tried to climb Mauna Loa from the west side. He failed both times. He then asked Hawaiians how to climb the volcano. They advised him to make the ascent in stages over several days to prevent altitude sickness and to follow a trail on the southeast side of the mountain. He took their advice, taking nine days to reach an elevation where ice formed at night. The next day he and his party were tramping through snow drifts and reached the summit the next day, February 16, 1794.

One of the most famous ascents of Mauna Loa was made a half century later by members of the Second United States Exploring Expedition led by United States Navy Lieutenant Charles Wilkes. Initially, Wilkes ignored the advice offered by Hawaiians. Instead, he set out from Hilo, reached the summit of Kilauea two days later and headed across country for the summit of Mauna Loa, guided by a midshipman who held to a straight course by consulting a compass. After two days, the expedition became stranded with no water or food on an open lava field. Wilkes finally sought the aid of Hawaiians who steered him directly west to a trail—the one Menzies had used—known as the Ainapo Trail. After four more days, Wilkes and his party reached the summit where they camped for twenty days, making measurements of barometric pressure and of the magnetic field. They erected tents that they surrounded with low stone walls to reduce the effect of the violent wind that blew constantly. Only during a few hours at midday did the temperature ever rise above freezing. One of Wilkes' officers who wrote an account of the ordeal remembered suffering terribly from the cold despite wearing a pea jacket, pantaloons, a wool scarf, a Scottish cap and two pairs of woolen stockings and spending most of the time huddled beneath several frocks and flannel blankets.

———•———

It was Friday, September 26, 1919, and Isabel was at home chatting with two guests—Mrs. Montague Cooke of Honolulu and Madame Madalah Masson, a concert pianist visiting from Australia. The three women were sitting beneath an awning on the verandah of the Jaggar house on the cliff edge of Kilauea drinking tea. It was a little before 6 P.M. and, much to their surprise, as Isabel recalled, her husband came "shooting out of the door doubled up like a jack-knife as if he had a sudden pain under his pinny." He straightened up as soon as he got out from under the awning and looked across the caldera, and then flew back through the house and up the steps that go up the cliff, all without saying a word. The three women looked at each other in amazement. Then Isabel stood and exclaimed: "Mauna Loa!"

Already a great column of fume, colored a burnt orange, was rising from the top of the mountain. She and the other two women raced up the steps and followed her husband.

By the time they got to the observatory, two fume clouds were rising high above the distant volcano. In the concrete cellar, the needle of the seismograph was vibrating wildly. The weather was clear that night and the whole sky above Mauna Loa was lit a brilliant apricot glow. By 3 A.M. the glow was gone and the seismic vibrations had stopped. The eruption had ended.

Nevertheless, the next two days were hectic. Thomas Jaggar had been planning to go again to the summit of Mauna Loa, and so he made preparations, whether the eruption resumed or not. He spent these days calling ranchers on the telephone asking what they saw, hunting up riding and pack animals and searching for guides. Isabel busied herself arranging camping outfits, getting out warm clothes and reinforcing boot soles with patches she cut from worn automobile tires.

Cooke and Masson were still at the Volcano House. This was Masson's first trip to the volcano, and so they had secured a room with a window that look toward Mauna Loa. Masson, in fact, was so impressed with what she had already seen that she spent most of the nighttime hours looking out the window. Early on the morning of Monday, September 29, she again saw a glow.

She woke Mrs. Cooke. The two of them hurried to the Jaggar house. The glow was now brighter than it had been three days earlier.

The Jaggars dressed and went to the observatory, followed by Masson and Cooke. The needle of the seismograph was again vibrating. Thomas Jaggar took a few photographs of the eruption. Next he headed to the Volcano House to call ranchers and ask what they saw. Meanwhile, Isabel packed a car with saddles, tents, blankets and food, enough to support six men for an entire week. When daylight came, her husband gave her instructions. She was to remain at the observatory and watch the lava lake within

Halema'uma'u and record if any changes happened. Then he and Finch and Lancaster left for the summit of Mauna Loa.

"All day the eye was busy," Isabel wrote of this day to a friend. But, the next day, after news arrived that a lava flow was coursing down the west side of the volcano and was about to cross a road, she "decided that the observatory's mechanic's eye was all right for Halemaumau." And she left to go and see the flow for herself.

She rode in a car with three others, including Dr. James Judd and his wife, Louise.* The drive took about three hours. They arrived at the active lava flow late that morning. By then, lava had already crossed the road.

"It moved so slowly that we stood right at the foot of it," Isabel recorded, "and watched its movement, pushing and tumbling on the grass, sometimes enough dropping down to leave a flaming, glowing oven."

Along the edge of the road the flow was about fifteen feet high and Isabel decided to scale it, noting that "though glowing between the chinks, [the lava flow] was cool enough on the surface to climb its ridge, although care had to be taken to avoid flaming holes that might easily set fire to the clothing."

When she reached the top she realized she was standing at the top of a levee of stationary lava, a hundred feet or so wide, and that there was an identical levee on the opposite side of the flow. In between was a broad stream of red-flowing lava, about four hundred feet wide, that was racing down slope, making "the sound of rushing water." Within the stream, as Isabel continued to observe, were "large bowlder-like masses of glowing matter coming down with the current—sometimes riding smoothly, sometimes rolling over and over, and sometimes breaking up into fragments." Whenever the last occur, "red lava spray would fly into the air." She walked up and down the crest of the levee to get different views, always keeping in motion to lessen the effect of the extreme heat.

---

* This is the same Dr. Judd who removed the brain tumor from Isabel's first husband, Guy Maydwell.

She returned the next day, again with the Judds. By then, the stream of lava had reached the sea, its entrance point marked by "a beautiful snowy white steam cloud going up thousands of feet."

She and the Judds followed a rough road down, driving to a place close to where the plume of white steam was rising. They met two young men who Isabel had known as boys at Holualoa School who directed her to a narrow path that ran along a steep rock coastline and that would lead to the plume. Isabel took the path. The Judds stayed behind.

The distance was not great, but she had to walk in front of a "crackling" hot wall of lava—the crackling sound indicating the lava was still moving, though slowly. The space between the hot wall and cliff edge was only ten or twelve feet. She then stood on a small rock promontory about a hundred feet above the sea and over the point where the steam was rising.

Here she "could see fragments of red lava falling into the water—rather into the steam, for the surface of the water was really not visible, the vapor so thick." Occasionally, an explosion threw rocks ten to thirty feet in the air. She would have liked to stay longer, but one of the young men called out that the hot wall of lava was shifting forward and might block her return.

She did return, and, within minutes, a large block did fall from the side of the wall, "leaving a great red glowing scar, like a furnace door left open." "I suppose it was well to be a little cautious," she later admitted.

She and the Judds drove back up to where the flow was streaming across the road and ascended the levee again, this time, the Judds joining her. "We stayed long enough to get a good view of it all after nightfall," she wrote to a friend, "though I can assure you there was no darkness in that vicinity. The light from the fiery river and the reflection from the arch of cloud above us where the column of vapor from the boiling sea made it much brighter than any moonlight night."

Her husband returned after four days on Mauna Loa. He had reached the eruption site, but, because of a mishap with glass collection tubes, had been unable to collect volcanic gases. He planned to return, but, before he did so, and after getting two days of rest, he decided to travel to the west side of the island with Isabel and see the lava flow that was crossing the road and pouring into the sea.

"I was terribly disappointed," Isabel wrote when she saw the flow activity again, "for as we approached there was no wonderful giant mushroom of steam to point out and when we got to the flow not a sign of a lava river could be seen." To add to the disappointment, her husband teased her, according to her, insisting "that all the wonderful things I had so excitedly told him about on his return from the source rift on the mountain side were imagined!"

The lava flow had, indeed, changed. The stream of molten lava was now only half as wide as during her previous visits, and the level of lava in the channel was much lower. Dark crusts of solidified lava rafting down the stream covered the flowing lava, and so red lava was difficult to see. And there was no sound.

Isabel did notice that a natural bridge had formed over a place where the stream was especially narrow. And she watched a man pass over the bridge. She decided to do the same.

"I, like a child, wanted to cross, too, and started down towards the bridge, much to [Tom's] disgust, for he had to pick up his tripod and camera and follow me—at least, he thought he had to."

She reached one end of the bridge and was starting to cross when the lava flowing in the channel suddenly started to rise. It eventually overflowed the levee. She took to higher ground, watching as the bridge was swamped by a surge of molten lava.

———•———

Back home at Kilauea, Jaggar decided to immediately reascend Mauna Loa and return to the eruption site, hoping to collect gas samples. So on Wednesday, October 8, the ninth day of the eruption, accompanied by James and Louise Judd and Isabel, they left the observatory and drove to a nearby ranch where they were met by two guides with riding and pack animals—and started up the mountain.

"We had a most interesting ride through the ranch lands," Isabel wrote, "sometimes through forests of large ohias, then across old lava flows, then through forests of great koa trees with lovely red thimble berries in abundance." There were cattle, wild goats and signs of wild pigs. By mid-afternoon they reached a water hole surrounded by tall grass and trees. Here they camped the first night, the fume column from the eruption in view, changing from brownish edges by day to rosy at night, the source about seven miles away. Occasionally, the silence of the night was disrupted by the sound of the distant fountains.

At daybreak, they loaded the animals and "started forth with a little water, some food, two cameras and the precious vacuum tubes." They rode until close to noon when they came to a steep, high wall of an old lava flow. Here they left the animals with some barley, and started on the hardest part of the journey.

For miles, they had to step from block to block over loose rough old lava or to stagger about, occasionally crashing up to their knees through the shelly surface of the smooth kind. Nowhere was a trace of soil to be seen. They crossed deep crevasses as much as five feet wide.

They climbed down and then scaled the high walls of old lava channels. Rain and a dense fog enveloped them. They now had to rely on Jaggar who had made this trek the previous week and who had his compass along. But he was bothered that no sound of fountains could be heard, though he had heard them distinctly from their location the previous week.

He stuck to his course in spite of the others suggesting different directions. Finally, as evening approached and the rain and fog cleared, there stood the active cone with a fume column rising from it.

The cone had grown considerably since the previous week. Its base was now more than a thousand feet wide and its top stood more than two hundred feet above the surrounding plain. They soon got within sight of the great ground fountain that pulsed from the center of the cone, each pulse sending out a barrage of molten stuff that hit the inner wall, then rolled down while still red.

The party, led by Jaggar, began an ascent. They trudged through pumice that became knee deep and excessively hot as they neared the summit. Once on top, they could look down into a horseshoe basin frothing with lava.

The air was filled with droplets of liquid lava. The sound was likened to giant surf breaking on a beach, only deafening. As Isabel recorded it, at times, there were "great explosions which made you feel that at any moment you might be sent hurtling into space and next morning your friends would find 'pieces of your mortal frame asticking to the sun.'"

From their perch high on the cone, they could see that the open end of the basin was a gorge, perhaps forty feet wide, where liquid lava was rushing like a torrent from the sluiceway of a dam. At first, the direction of the current was straight, then, after a few hundred yards, it bent and plunged over a fall into a brightly glowing abyss. And beyond, looking out to the west, in the direction where lava was flowing into the sea fifteen miles away, was a desolate country covered with glowing and flowing lava fields.

By the time they descended, it was night, and the others ate dinner while Jaggar went in search of gas samples. He found a convenient place near the sluiceway where there was a wide glowing crack a few yards from flowing lava. He had eight tubes. He successfully collected gas in two of them.

Back at camp, he and the others began to settle down for the night, but were soon disturbed by a slow moving lava flow. "We were sorry to move away," Isabel wrote of that night, "but we eventually had to do so as the flow that had hurried us down from the side of the big cone and which continued advancing all night, began to spread and spill over its sides toward our nice comfortable bivouac."

Another place was found nearby on higher ground. Here they spent the remainder of the night. "Some of the party got a few winks at intervals but with no blankets above and hard lava beneath, I think their slumbers must have been very light."

Isabel's boots were still wet inside from the day's rain, and so, "with cold feet below me and hot roaring lava fountains in front of me and a brilliant moon taking on all sorts of complexions as it passed through the volcanic smoke above me, it was useless to try to sleep. So I just sat and looked. And what wonders I witnessed!"

It was still hours before the sun rose, and, when it did, "what changes in sky, in clouds, in mountain, and especially in the magnificent pillar of smoke directly opposite the rising sun and with the fire fountains at its foot!"

She ended by writing, "I cannot think of it without tears welling up. I have never seen anything like it, and I have seen some wonderful sunsets down in the crater at Kilauea."

———•———

The 1919 eruption of Mauna Loa was a hallmark in the study of Hawaiian volcanoes. The gas samples Jaggar collected—one admirer later lauded him for "using both his knowledge of crater activity and a very real courage"—are the best ever collected for an eruption of that volcano, even to this day. An analysis of the samples by Shepherd and Day at the Carnegie Institution of Washington supported their claim, made a few years earlier, for Kilauea: Volcanic gases consist primarily of water vapor and a significant amount of carbon dioxide.

Jaggar was also able to document, in detail, a typical Mauna Loa eruption, which began with an outbreak in the summit, lasted several hours and was characterized by fiery fountains, then shifted to a lower elevation where a voluminous outpouring of lava took place. In 1919, it took about a day for molten lava to reach the sea—a forewarning of what was to come.

———•———

In early March 1926 stories started to circulate that people were having encounters with the volcano goddess Pele. "A queer old lady, agile as a nymph" was how one newspaper report described her. And, for some unknown reason, she was spending her time stopping vehicles along highways, hoping for a ride, or walking up to remote houses and asking for food. The reason behind the sudden flurry of sightings, according to some people, was a comment made by Jaggar.

He had recently addressed a crowd of tourists. When asked what was the likelihood that an eruption would happen soon, he purportedly answered, "Pele might return tomorrow, perhaps, next week, or maybe next month." Later, when asked about his comment, he said that his statement had been "overly stressed" by a newspaper reporter in Honolulu. In fact, he declared, there was no positive means to predict when volcanic activity might resume. "Lava may return to Halemaumau quietly and without warning." And it may be preceded or followed by an eruption of Mauna Loa.

But the stories persisted. On April 7, 1926, a report on the front page of the Hilo newspaper said, "Hawaiians believe that something unusual is about to happen." That afternoon it did.

Hope Carlsmith and her husband, Leonard, had arrived in Hilo a few days earlier to visit his parents. On the afternoon of April 7, the four of them took a trip to Hilo Bay to swim at the local yacht club. The four of them were floating on a raft when Hope said that she was going to swim to shore and back. She dived into the water and was about twenty-five yards from shore when a shark attacked her.*

Her husband saw the attack. He saw a swirl around his wife, then the fins of the shark. He saw the shark grab hold of his wife and pull her under. Fortunately, the shark let loose of her. Her husband dived into the water.

---

* Shark attacks are rare in the Hawaiian Islands. This was the second known shark attack in Hilo Bay and only the fifth known shark attack anywhere in the Hawaiian Islands.

He swam to his wife and pulled her to shore. There he could see that the shark had lacerated her right leg from the heel to the thigh. The calf of her other leg was torn nearly to shreds.

Hope was losing a great deal of blood. Someone on the beach had a necktie, and Leonard used it as a tourniquet, tying it to her upper right leg. A doctor said later that this had saved her life.

Three days later, while lying quietly in a hospital room, Hope was disturbed when her husband and in-laws rushed in. They had come to comfort her. They said that Mauna Loa had started to erupt and that some people in Hilo were saying that her shark attack was the event that not only foretold of an eruption, but that the coming eruption would be a destructive one.

Then, before leaving, her husband rearranged her room, moving a large mirror so that Hope could see the reflection of the red glow of the sky from her hospital bed.

———•———

Unlike at most volcanic sites in the world, on the island of Hawaii, the reaction was one of celebration. The flow of lava into the sea in 1919 had brought hundreds of tourists to the island. This eruption, which had its source at the same place on the volcano as the previous two, might prove to be an even bigger bonanza.

On the second day of the eruption, a full page of advertisements appeared under the headline banner: WELCOME HOME TO MADAME PELE. And beneath that was written: More Lava, More Tourists—More Tourists, Better Times.

The advertisements were introduced by a quote from Jaggar who had said "that he expected this flow would be greater than that of the 1919 flow." One advertisement reminded volcano watchers to buy flashlights, flashlight batteries, photographic film and cameras. Another one claimed "Fuller's paints are just as bright as the volcano's fire and flow."

By the third day, the celebration was still going strong. A front-page headline proclaimed: The Volcano is Erupting in Earnest

Now—It is a Time of Rejoicing All Over the Isle." That night, 843 guests registered at the Volcano House, the second largest ever. The hotel had beds for 200.

On the first day of the eruption the Interisland Steamship Navigation Company announced that it would be sending a ship daily from Honolulu to the island of Hawaii for those who wanted to watch lava flow down the mountain and toward the sea. On the second day, the United States Navy sent two seaplanes to take aerial photographs of the advancing lava flows. On the third day, the United States Army Air Service sent three planes to repeat the photography.

And on the fifth day a report came to the observatory that an advancing lava flow was about to cross the road. This flow, according to the aerial photographers, was coursing its way between the now long-congealed 1916 and 1919 lava flows and was headed to the sea.

Thomas and Isabel Jaggar left immediately for the west side of the island. They arrived at half past ten the morning of April 16. The active lava flow had not yet crossed the road. A crowd of people had already gathered to watch the event.

One of those who came to watch later described the mad dash to see the flowing lava as akin to a rush toward a new gold field. Hundreds of cars seemed to pour down the roads from elsewhere on the island and into south Kona, causing the biggest traffic jam the island had yet seen. Many cars ended up stalled by punctured tires or overheated engines, their occupants picked up by others who were racing to the lava flow.

When the Jaggars arrived, the flow was still on the uphill side of the road. Smoke of the burning forest was just visible from the road. A short trek showed that the front of the flow was about a thousand feet wide and as much as thirty feet high. It was creeping its way through a dense tangle of guava and lantana bushes.

About noon the front of the flow surged. It could be heard breaking through bushes and crashing trees to the ground. It was soon "seen coming on like a great fiery dragon of red hot stones, pushing forward several feet a minute with a continual tumble of black fragments at its front."

A few individuals rushed forward to have their photographs taken in front of it, standing motionless only a few seconds as waves of intense heat and noxious vapor swept over them. As the front neared the centerline of the highway, people from opposites sides ran out to shake hands and to wave a final goodbye. A few ran up to red-glowing blocks that had tumbled down from the flow to burn paper or singe a hat or a wooden walking stick.

Both the home of local fisherman George Ka'ana'ana and the Catholic Church stood within a few hundred yards of the advancing flow. Mrs. Ka'ana'ana set red handkerchiefs at the corner of her property, then walked close to the flow front where she placed offerings of sugar cane and sweet potatoes with the vines still attached. The Catholic priest, Father Eugene, and several friends of the Ka'ana'ana family urged them to save what they could from their house and tried to convince them that the offerings would not halt the lava flow. But George Ka'ana'ana refused, saying that as long as no one touched him or his possessions, Pele would not harm him. His wife then came out with a set of white teacups and offered a drink of water to anyone from the family's water tank.

But the temptation was too great. Father Eugene and a few others rushed into the Ka'ana'ana house and dragged out what they could, filling a waiting truck with furniture and other items. Soon after, the Catholic Church, which stood nearby, was touched by lava and burst into flames. Mr. and Mrs. Ka'ana'ana were sitting inside their house when a small lobe of molten lava reached out from the main flow and caused their house to go up in flames. The Ka'ana'anas quietly left.

After the Catholic Church and the Ka'ana'ana house burned and the smoldering ruins were toppled and covered by the advancing lava, the flow seemed to slow. Some onlookers thought it had stopped. It was still three miles to the ocean, all of it down a steep slope covered by a dense forest. And at the bottom was the fishing village of Ho'opuloa.

The Jaggars slept that night at the single store in Ho'opuloa where, just before their arrival, the owner had removed his stock of goods and furniture, then swept the building, leaving the floor spotless and clean.

Much of the night was a quiet one. A red glow, of course, could be seen high above them on the steep slope. But whether the flow was still advancing was unknown until three o'clock when Thomas Jaggar looked up and saw the first faint red glows of the lava flow starting to pour over the top of the bluff back of Ho'opuloa.

By sunrise, the flow was well down and coming straight for the village. More than a hundred people were now at Ho'opuloa, most of them having slept in their cars during the night. As the sun rose, many joined in a prayer service.

Jaggar went through the crowd warning people to stay away from the shoreline as the hot lava would explode when it came into contact with cold seawater and that those same explosions might produce large sea waves. He also asked that everyone keep a sharp lookout for any strange fish that might be brought to the ocean surface after the lava entered the water. Anyone who found one was asked to preserve it immediately.

Jaggar also went through the village warning people that this was their last chance to save possessions. That evening, trucks arrived with men who had been sent by the county government to assist, if asked, in removing lumber or other reusable material from buildings. Only three families asked that their houses be disassembled. The other dozen or so families refused the aid.

That evening the steamer *Haleakala* of the Interisland Steamship Navigation Company arrived and stationed itself a mile offshore. Its three hundred passengers gathered aft to watch the spectacle that was happening on shore. Many reclined in steamer chairs and were served drinks, while others sat on benches or atop beams or wherever they could find a place to sit. Even from the shore, the ship's orchestra could be heard playing Hawaiian music.

From the ship's deck, passengers could see the entire length of the lava flow as it coursed its way down the mountain. One passenger would describe the reflection off the calm sea as "a flaming ocean such as has never before been photographed."

Just before midnight, passengers could see a bright flare shoot up from a spot high on Mauna Loa. It was bright enough to illuminate the area for many miles. Immediately after, again as one of the passengers would describe it, a surge of lava broke "forth from near the source of activity high up in the mountain and [went] rolling and tumbling down the side, a great seething river of fire, carrying everything in its wake and completely covering all traces of the former flow except along a line near the sea."

Everyone on board was tense with excitement and was certain that the surge would surely send lava into the sea. But it stopped just short, though every few minutes a new line of lava near the coast brightened suddenly, causing the passengers and the ship's crew to shout with excitement.

Those who were at Ho'opuloa also saw the surge just before midnight, though they had no idea what was happening elsewhere. To them, it "resembled a great ball of fire tumbling down." And, with that, everyone left Ho'opuloa, including the Jaggars, who retreated south along the coastline.

At four o'clock the first house in Ho'opuloa caught fire. At six thirty, the store where the Jaggars had stayed and slept the night before began to burn, the moving lava pushing it sideways as the building burned.

Minutes later, the first red incandescent boulder rolled off the flow front and fell into the sea. The Jaggars stood nearby and watched as the flow itself entered the water, causing "clouds of steam peppered with black sand" to shoot up. Within an hour, Ho'opuloa was covered by lava. Army airplanes flew overhead the entire day photographing the flow of lava into the sea.

That night, the flow of lava toward Ho'opuloa stopped and a new lava flow started to pour down the east side of the island. But this one was short-lived. Within a few days, the lava stopped moving. And the 1926 eruption of Mauna Loa was over.

One of the passengers on the *Haleakala* described what he had seen as "a big performance." One of the pilots who flew over

Ho'opuloa on its last day of existence said he was "struck by the spectacular beauty spread out before him."

A reporter for the Honolulu *Advertiser* who filed several stories about the eruption—and who was sympathetic to the Hawaiians who had lost their homes and who believed in Pele—ended his last story by concluding the eruption had been "a vivid lesson in geology but a greater lesson in religion."

Later, some Hawaiians would say that Ho'opuloa would have been saved from the lava flow if strangers had not intervened.

CHAPTER FOURTEEN

THE GODDESS

Mauna Loa began to erupt on November 5, 1880, sending a broad stream of lava down toward Hilo and its bay. By March the flow front was within seven miles of the bay. By early June, it was within five miles. Crowds of people now made the short trip to see the advancing lava. One person who spent a night camped close to the edge of the slowly moving lava wrote, "The sight was grand. The whole frontage was one mass of liquid lava carrying on its surface huge cakes of partially cooled lava." The lava continued to advance. By the third week of June, the lightheartedness ended.

On June 26 the flow entered a narrow stream channel that caused it to surge forward and pick up speed. After another week, the front had advanced nearly a mile. People were now concerned that the town of Hilo would soon be lost.

Frederick Lyman, one of Hilo's leading citizens and the representative of the Royal Governor of the island of Hawaii—the governor resided in Honolulu—proclaimed July 6, a Wednesday, to be a day devoted to prayer. All businesses were closed. During the morning, local ministers led church services, asking God to halt the advancing lava. That afternoon, the ministers and Lyman took the short walk to see the lava. During the few minutes they stood there, the flow front advanced twenty feet.

By July 18 the fate of Hilo seemed to be sealed. Lava was now within a mile of the nearest building and less than two miles from the shoreline. People expected their houses and stores to soon be covered by lava. The same day several local Hawaiians sailed for Honolulu where they sought a meeting with Ruth Ke'elikolani, a sister of King David Kalakaua who was then on a world tour and who had left Ke'elikolani in charge of the government. The visitors asked Ke'elikolani if she would come to Hilo and see for herself the disaster that was about to befall their town. She agreed to go, but she did not travel straight to Hilo.

Instead, she chartered a steamer that transported herself and a small retinue of advisers to Kailua on the west side of Hawaii. From there, she and her advisers took five days to travel by wagon and on horseback to Hilo, arriving on August 4.

On the day they arrived, the people of Hilo were holding a meeting to discuss what might be done to save the town. Plans were made to construct earth embankments around important buildings and a stone wall around a local sugar mill to divert the lava. It was also proposed to use dynamite to disrupt the flow near the eruption site, though nothing came of this suggestion. Ke'elikolani did not attend the meeting. When she arrived in Hilo, she took up residence in the house of a close friend where she remained for five days.

On the morning of August 9, she sent for one of her advisers. She told him to purchase all the white silk scarves that could be found in Hilo. She also asked that a bottle of brandy be purchased to substitute for *awa*, a drink common throughout Polynesia, including the Hawaiian Islands, and which had a mild sedative quality. After the scarves and the brandy were brought to her, she asked for a wagon to carry her to the flow front, but, because of her great weight—she weighed over four hundred pounds—the wagon broke and a sturdier one was found.

At the flow front, Ke'elikolani walked alone, all onlookers, both Hawaiian and non-Hawaiian, keeping a respectful distance. She was seen sprinkling drops of brandy and placing scarves on the red lava. Some people thought she spoke a few words. Then, as one of the onlooker recorded, she left the lava flow and Hilo "with confidence and returned to Honolulu."

The next morning when the people of Hilo looked up at Mauna Loa they saw that the eruption had ended. "A remarkable coincidence," explained the whites. "The work of Pele," whispered the Hawaiians, even though the last temples dedicated to the goddess had been destroyed sixty years earlier.

————•————

Pele is an *akua malihini*, a foreign goddess. She arrived in the Hawaiian Islands in about the 14th century, hundreds of years after the first humans arrived. According to an ancient chant, she came from *Kahiki*, probably Tahiti, "from the land of Pola-pola," which might be the island of Bora-bora.

When she arrived on the island of Hawaii, according to another chant, she was told that a god, Aila'au, already inhabited the volcano of Kilauea. Wishing to meet him, she started up the mountain. But, when Aila'au learned that she was approaching, he left. Ever since, Pele and her family, which included many members who represent other elemental forces, such as earthquakes and storms, have resided on the island.

Her importance and wide veneration is indicated today by her association with more than a hundred place names found on all the major islands and on the small leeward ones. Many of these places are of obvious volcanic origin. For example, Koko Crater on the southeastern shore of Oahu—a cone of volcanic ash that formed in prehistory when lava erupted through shallow seawater—is similar in form to female genitals, which explains why the ancient name of the crater is *Kohele-pele-pe*, Pele's vagina.

One of the early compilers of Pele stories was Abraham Fornander, who came to the islands in 1841 on a whaling ship, then deserted. He rose to prominence in island society, first as a newspaperman, then as a judge. Fornander became interested in ancient stories because he thought they might contain evidence that Hawaiians were a lost Aryan tribe, a theory then popular among some westerners. No such connection exists. The stories he recorded contain some of the most vivid depictions of volcanic activity in Hawaiian lore.

One tells of a battle between Pele and the pig god, Kamapua'a. Pele sees Kamapua'a in the crater Halema'uma'u, and so she starts an eruption. According to Fornander, Kamapua'a, who is usually in the form of a young man, "changed himself into the form of a giant hog, opened its mouth, showing its tusks, and swallowed Halema'uma'u, taking in Pele and her sisters." The battle continues as Pele chases him through the forest.

Such a collapse, or swallowing, of the Halema'uma'u crater followed by the flow of lava—represented by Pele chasing Kamapua'a through the forest—is a familiar sequence. Jaggar recorded such a sequence in 1919, 1922 and 1923. And the sequence has occurred more than a dozen times in recent decades. As the Pele-Kamapua'a story attests, it was also seen and recorded by ancient Hawaiians.

Another story tells of a confrontation between Pele and one of her sisters, Hi'iaka. In this one, compiled by Nathaniel Emerson, who had stories that had been published in Hawaiian-language newspapers translated into English, Pele has fallen in love with a young chief, Lohi'au, who lives on the distant island of Kauai. She sends Hi'iaka to fetch him, promising her sister that she will not destroy Hi'iaka's favorite forest during her absence. But Hi'iaka is delayed. As she returns and nears the island, she sees that her beloved forest is on fire. Understandably upset, Hi'iaka confronts Pele who rages by killing Lohi'au and throwing his body into

Halema'uma'u. Hi'iaka then digs furiously to recover the body. As Emerson recorded the story, "She tore her way with renewed energy; rock smote against rock and the air was full of flying debris."

This story probably records two volcanic events witnessed by Hawaiians—and which can be seen in the geology of Kilauea. The first is a decades-long eruption during the 15th century that burned and covered more than a hundred square miles of dense forest with lava. The second is the subsequent collapse of the summit of the Kilauea to form the caldera, an event that was accompanied by explosions, the "flying debris" tossed by Hi'iaka as she dug for Lohi'au.

Throughout most of the 19th century, there was a general disregard for any public display of reverence for Pele—or for anything related to ancient Hawaiian culture. Not until the 1870s was there a renaissance of Hawaiian culture, brought on by King David Kalakaua who was trying to establish an identity for the islands separate from the one then dominated by American missionaries. Ancient chants were heard again. Public performances of hula were staged, but still denounced as scandalous by some non-Hawaiians. Ke'elikolani's trip to Hilo in 1881 was one of those public renewals of faith in Pele—one that illustrates a conflict between Hawaiian and western views of the volcanoes.*

Through western eyes, a volcano is a physical system that can be understood and, possibly, controlled. To Hawaiians, the volcano is an important part of cultural identity as represented by Pele, and that is understood through analogies. For example, the redness of molten lava is also seen in the redness of the pom-pom flowers of the ohi'a tree and the berries of the ohelo bush, related to cranberry, that grow on Kilauea. When blood-like, red-glowing

---

* Even though public displays of Hawaiian culture were discouraged in the 19th century, private ones were still conducted, including those that paid homage to Pele. During such private ceremonies, offerings would be made to Pele, which included fish, stalks of sugar cane, taro and flowered leis. During especially important times, the sacrifice of a pig was done.

lava erupts and flows toward the sea, it is the menstrual blood of the goddess Pele, an act that mimics the march, in ancient times, of Hawaiian women who cleansed themselves near the ocean during menstruation. To alter the path of flowing lava is to disrupt a woman's menstruation, akin to disrupting nature herself.

----·----

On the evening of November 21, 1935, after a day of earthquakes, some felt as far away as Honolulu, lava began gushing from a fissure high on Mauna Loa. The next morning an airplane was sent by the commander of the United States Army Air Corps at Wheeler Airfield on Oahu to take Jaggar on a flight to see the eruption.

The flight lasted an hour and ten minutes. The pilot flew low and circled as close as possible to the eruption site, the airplane shaking violently in the highly turbulent air rising from the hot lava.

Beneath him, Jaggar could see a wall of fountains a thousand feet long. Molten rock shot up hundreds of feet. Five ribbons of liquid lava were coursing down the north side of volcano as braided streams. Later, back on the ground, Jaggar could see that the ribbons had coalesced into one broad river. By nightfall, the lava had flowed down the steepest part of the volcano and the front was on the broad plain between Mauna Loa and Mauna Kea. The question now, after only one day of activity, was whether the eruption would continue long enough for the front of the flow to run up against the rising slope of Mauna Kea and turn either to the west and onto vacant land or to the east and toward Hilo.

The critical point came in late December. For four weeks, the flow had crept over flat ground. The front itself was a continuous wall of stiff lava that acted like a cofferdam, holding back a vast pool of liquid lava. As the eruption continued, the flow front did move forward, the progress measured in a few feet a day, though more importantly the volume of the lava pool continued to increase.

On December 22 the flow front reached the midpoint between Mauna Loa and Mauna Kea. To continue northward the front would have to move uphill. And so, instead, the flow began to spread both to the east and the west. The westward movement was short lived, halted when that edge of the flow pushed up against the side of an older lava flow, one erupted in 1843. Now there was only one direction for the lava to flow—toward Hilo.

On December 23 liquid lava did break out and surge to the east, the flow front advancing more than a mile that day. On December 24 it moved another mile and a half. At those rates, lava would flow over Hilo in early January. Jaggar decided that something had to be done, and it had to be done fast.

Years earlier, at Kilauea, he had watched as a lava flow changed course after a section of the channel the lava was flowing in collapsed. He reasoned that a lava flow could be redirected by controlling where such collapses occurred. But how could the edge of a lava flow be caused to collapse?

Thurston suggested the use of explosives. His idea was to erect a wooden stand close to the point where one wanted to disrupt the flow, then, with a remote mechanism, drop TNT explosives from the top of the stand. There were obvious practical problems with the idea, such as how to keep the wooden stand from catching fire, but Jaggar thought the idea a good start.

Guido Giacometti (then the supervisor of the Olaa Sugar Mill and a man of practical experience, having graduated from the Polytechnic Institute in Zurich, Switzerland, in 1901, the year after Albert Einstein had graduated from the same institute) had a solution. The precision of aerial bombing had improved greatly since the end of the First World War. It might provide the means to deliver explosives at the edge of a river of lava. "The more I thought of Mr. Giacometti's suggestion," Jaggar decided, "the more I realized that that would be the feasible thing to do."

Jaggar sent a radio message to the Army's commanding general on Oahu describing Giacometti's suggestion. The same afternoon, an airplane with three officers from a bombing squadron flew over Mauna Loa to see the eruption. At noon the next day, a conference was held in Hilo attended by Jaggar, several local officials and officers of the Army. They issued a joint statement that said: Because the public "demands that something be done to divert or stop this lava flow," they recommend the dropping of bombs. The statement was sent to the Army's commanding general who issued orders for the immediate shipment of the needed bombs to Hilo. The request apparently reached all the way to Washington, D.C., because the next day a telegram arrived for Jaggar from President Franklin Roosevelt: "Planes authorized. Good luck on your effort."

Fourteen Army planes from Luke and Wheeler airfields on Oahu arrived in Hilo on the morning of December 26 to carry out the bombing of Mauna Loa. The planes included ten Keystone bombers, two amphibious planes and two observation planes, the last to carry spotters and photographers. Twenty officers and nearly forty enlisted men also arrived. An Army transport ship with the bombs reached Hilo that night.

Work began the next day to locate targets. Jaggar flew in one of the observation planes with Colonel Delos Emmons, who was in charge of the operation. They decided the primary target would be at the edge of the lava channel near the eruption site high on the volcano. A secondary target would be about a mile farther down.

That night Thomas and Isabel Jaggar and several Army officers and enlisted men climbed to the top of a cinder cone on the south side of Mauna Kea where they would watch the bombing. Jaggar would talk directly over a short-wave radio with people in Hilo.

The next morning the weather was perfect, neither wind nor clouds. The first bomber took off at 8:45 from Hilo, followed by four others at twenty-minute intervals. Because the planes were armed with bombs, the pilots were ordered to avoid flying over Hilo. And so they assembled over Hilo Bay, reaching an altitude of twelve thousand feet, then turned toward Mauna Loa.

In turn, each plane pitched forward and dove at a target, releasing a pair of bombs. As seen by Jaggar ten miles away, when the first bomb hit, it caused a flash, followed a minute later by a deep boom. Minutes later, a second flash followed by a weak column of smoke rising from the impact point. Seven more hits within the next hour, the last one producing a double puff of smoke as two bombs hit together. Then the bombers returned to Hilo.

A second set of planes left Hilo that afternoon and released their bombs over the secondary target. Then the two observation planes flew over the lava flow the remainder of the day, the crews noting the progress of the flow front by using trees as markers.

"Our purpose was not to stop the lava flow, but to start it all over again at the source so that it will take new course," explained Jaggar the next day to newspaper reporters. And it seemed to have that effect.

On the day before the bombing, the flow was advancing as fast as 800 feet an hour. Hours after the first bombs were dropped, the rate had dropped to 150 feet an hour. And late on December 28, the day after the bombing, the flow front had stopped.

The experiment seemed to have been a success, but there was also a lessening of the outpouring of lava on the day before the bombing. The last molten lava erupted on January 2.

Furthermore, a later inspection of the bombsites and explosions on the ground showed that few of the bombs had hit the intended targets. Though Jaggar claimed the experiment had been "entirely successful," today most people conclude that the bombing had no effect on stopping the advancing lava flow.

It was, as one non-Hawaiian wrote who witnessed the bombing, "like the time in 1881, when Princess Ruth Ke'elikolani tried to stop a lava flow from invading Hilo by throwing silk handkerchiefs and a bottle of brandy into the molten flow." In other words, in the view of that particular person, it had been a coincidence.

———•———

Though the roar of the explosions could not be heard in Hilo, those who were in town could clearly see small columns of smoke rising one after the other, the plumes blown steadily away from Hilo by the prevailing trade winds. There was no denying that the lava flow of Pele was under attack.

"Nothing good will ever come from disturbing Pele," said seventy-one-year-old Mr. Ebenezer Low, a local rancher and part-Hawaiian, the day after the bombing. "The Hawaiians don't like it, at least the old folks don't."

One of the pilots, William Capp, would recall, "None of the [bomber] crews went into town once we found out that [residents] were threatening the crew chiefs and ground support." As soon as the second wave of bombers returned to Hilo, the planes were refueled and flown back to Oahu.

Delos Emmons, who had flown with Jaggar on the initial reconnaissance flight and had decided on the primary and secondary targets, told an aide years later: "I will always remember it as it was like flying inside a fiery furnace. I was sitting on top of the gas tank expecting it to explode any minute. I was scared as hell."

"Didn't you know of the legend that anyone who interferes with the Goddess Pele when she is erupting must meet a violent death?" the aide asked.

"Sure I did," Emmons answered. "Everyone knows that and I did something about it, too. After we came in from the run, I made another with a passenger, a small pig that I bought. When I was over the center of the crater I dumped him out. It was the meanest thing I ever did and I can still remember the poor devil's screams as he catapulted below."

"Did the other flyers do likewise?"

"No. They gave me the horse laugh."

———•———

Less than a month after the eruption ended, two of the planes involved in the bombing of Mauna Loa collided over Luke Airfield on Ford Island in Pearl Harbor. It was the worst airplane crash to date in the Hawaiian Islands.

The two planes were among the last elements of three in a flight formation of nine bombers returning to Luke Field after a brief night flight. The three were flying en echelon, which means two were flying above and slightly behind the leader. When the planes were about a thousand feet above the ground, the two planes that were following touched wingtips. Both burst into flames immediately.

There were two distinct explosions followed by a mass of metal wreckage falling onto the field. Only two of the eight men in the planes survived.

One survivor was Lieutenant Charles Fisher. "I was piloting one of the planes roaring over Luke Field with both motors wide open," the lieutenant told investigators. "Suddenly there was a terrific impact. I turned to see a mass of flames in the rear of the plane and heard a violent explosion."

He unfastened his seat belt, then stood in his seat and dove through a hole in the wrecked plane. He remembered passing through the still-whirling propellers. "I don't know how they missed," he said. He pulled his ripcord. No sooner had he done that than he felt a tremendous jerk. His parachute, still unopened, had snagged on the steel ladder of an oil tank, leaving him hanging a few feet from the ground. He unfastened the chute, dropped to the ground, and crawled on all fours away from the tank.

The other survivor was Private Thomas Lanigan who was in the other plane. "We were flying nicely in a routine flight when the crash came. It knocked me to the floor," he said. He straightened up to see that the canopy was gone. He jumped and his parachute opened. He descended rapidly, heading straight for the flaming wreckage on the ground. At the last moment a burst of hot air from the wreckage tossed him upward, then the wind carried him sideways. He landed about twenty-five feet from the wrecked planes.

Of the eight men in the two planes, only Fisher and Lanigan had not been involved in the bombing of Mauna Loa. (Emmons, who sacrificed the poor pig, was not involved in the accident.) The other six were. One of those was Private John Hartman who was seen making a clean jump from his airplane, but, as he settled closer to the ground, a gust of wind came up and blew him into the fire.

An Army board of inquiry met to investigate the accident. The board concluded that the mid-air collision of the two bombers over Luke Field had been due to "extenuating circumstances." There was no evidence of mechanical defects nor was either pilot guilty of any apparent neglect. Instead, the accident had been caused by the slipstream of air coming off the lead bomber forcing one of the trailing bombers to lurch sideways and collide with the other.

Ebenezer Low and some other Hawaiians saw it differently. The six dead men had been victims of Pele's wrath.

And the story does not end there. The remains of the six dead airmen, which, according to the autopsies, where "burned beyond all recognition," were loaded onto an Army transport ship and sent to the United States mainland for internment. At the last moment, before the transport left, a seventh body was added. It was the mortal remains of Father Damien, the priest who had run the leper colony on Molokai.* Now, forty-seven years after his death, his remains, sealed inside a coffin made of koa wood, were being transported back to Belgium, the country of his birth. (The bodies of the six dead airmen were in metal caskets.)

The transport ship left Honolulu on February 4 and arrived six days later in San Francisco. Six hours before it docked, a mystery arose that has never been solved. The ship's captain, Edgar McClellan, disappeared. Members of a board of inquiry admitted that they were baffled. One theory held that the captain had committed suicide, but no traces and no notes related to a suicide were discovered. The board closed the investigation stating that it was unable to reach a conclusion as to the fate of the captain.

---

* For sixteen years Father Damien cared for the physical, spiritual and emotional needs of those in the leper colony. Eventually he contracted and died of the disease. He was canonized by Pope Benedict XVI on Sunday, October 11, 2009.

CHAPTER FIFTEEN

THE LAST VOLCANO

A s Jaggar was fond of telling visitors, the work of a volcano observatory does not start and end with eruptions. There is much else to do. There is the constant readjustment of scientific instruments and the reading of those instruments to decide whether an eruption is imminent. There is the need to prepare better topographic maps, ones that showed more detail, so that the paths of future lava flows might be predicted and so that remote locations may more easily be accessed. And there was a mountain of correspondence that needed to be answered, inquiries about the progress of the science and requests for copies of measurements already completed. There was the need to encourage other scientists to visit and apply their own expertise to a study of the volcano, whether it be a new way to collect and analyze volcanic gases and changes in gravity or magnetic fields or a host of other possible investigations, each one intended to reveal something new about what was happening inside the volcano. And there was the need for Jaggar to develop his own new techniques and invent his own new equipment and to extend his work to other volcanoes. Of this last effort, Jaggar got a surprising boost after the unexpected death of President Warren G. Harding.

Most medical experts who have studied the case have concluded that Harding's sudden death at the Palace Hotel in San Francisco on August 2, 1922, was due as much to a misdiagnosed heart attack of the previous week, which went untreated, as to an increase of personal stress after the recent revelation of financial scandals within the administration. The most famous scandal involved several key government officials—all close personal friends of the president—and the leasing of oil reserves in various places, including Teapot Dome, Wyoming, from which the scandal would get its name. The leasing had been done by the Department of Interior, and so, when Harding's successor, Calvin Coolidge, took office, he reacted to the scandal by reorganizing the Department and infusing it with new programs. One of those programs was the study of volcanoes, which the Chief Geologist of the United States Geological Survey, Walter Mendenhall, had been proposing for years.* It is through such convoluted politics and historical accidents that the fate of others is determined. For Jaggar, it meant that his dream of a national program to study volcanoes had finally been realized.

On July 1, 1924, the Hawaiian Volcano Observatory was transferred from the United States Weather Bureau to the United States Geological Survey. Jaggar immediately hired a mechanic, a machinist and a draftsman who would also work as a clerk. Finch, who now had four years of experience at Kilauea, was moved to

---

* Mendenhall and Jaggar had met in 1896 when both men were doing graduate studies at Harvard. Mendenhall's interest in Alaska began two years later when he was attached to a military expedition that explored a vast area east of Cook Inlet. Ever since, he was anxious for the United States Geological Survey to support someone to work in Alaska.

the small community of Mineral, California, where he started the nation's second volcano observatory at Lassen Peak. The intention was for Finch to expand his work to include all the volcanoes along the west coast, including Mount Rainier and Crater Lake, both already national parks, and a small mound known as Mount St. Helens. Meanwhile, Jaggar would continue to focus his attention on Hawaiian volcanoes and include those of Alaska.

Having a mandate to now work in Alaska, Jaggar reflected on his experiences in 1907, especially, the disappointments. The greatest one had been when the schooner *Lydia* had sailed close to a volcanic island, then, because there was no suitable place to anchor and reach shore, Jaggar had been forced to sail away, unable to set foot on the island. What was needed was a new type of vehicle that could negotiate ocean swells, then run up and onto a beach. Jaggar had been designing such a vehicle for years in his head. Now he had the chance to build and test one.

He returned to Alaska in 1927, traveling through Seattle where he bought a new Ford Model-T. He had the metal body removed and the transmission replaced with one that had extra low gears. Then, with the help of Ford engineers, he redesigned the rear axle to accept four balloon tires. He had the modified vehicle loaded onto a ship and sailed to Alaska.

Isabel was traveling with him. At Kodiak Island they separated. She spent the next two months traveling through the Yukon and the interior of Alaska. Her husband stayed a week on Kodiak, testing his vehicle by driving it up and down the beach, much to the amusement of cannery workers and their bosses.

From Kodiak, he found a ride for himself and his vehicle on a local steamer, the *Starr*, formerly a halibut schooner, the *Starr* had been converted so that it now carried passengers and mail once a month from Kodiak to small ports along the Alaskan Peninsula and Aleutian Islands. The passenger quarters were cramped, and the ship's motion was often violent. Thirty-six bunks were available, arranged in twelve stacks of three. To each bunk was attached an iron bracket that held a large pasteboard cup, in which most passengers rendered up at least a portion of their latest meal. As it was commonly said, few people ever sailed on the *Starr* for pleasure.

It was a three-day trip from Kodiak Island to the canneries at King Cove, where a motorized vehicle had never been seen before. And the testing was more rigorous. Jaggar ran the Model-T chassis over sandy beach, grassy flats and hard tundra. He dashed it through the surf. A local mechanic helped him attach a winch spool and the two men tested the vehicle to see over what surfaces and up which slopes the Ford motor had the power to pull itself. After a few days of such tests, Jaggar asked that the vehicle be sent back to Kodiak and shipped to the Hawaiian Islands where he would conduct more tests and make more modifications. In the meantime, for the remainder of the summer, he would see what he could of Alaska's volcanoes.

At King Cove, he was introduced to two local men—John Gardiner and Peter Yatchmenoff—who were hunting bears for a museum on the east coast. Gardiner owned a motorized skiff and he took Yatchmenoff and Jaggar thirty miles to Pavlof Bay.

At the entrance to the bay, the three men camped in a sod hut. To the north were twin snowy volcanic cones, Pavlof and Pavlof Sister. They planned an ascent of Pavlof, the more active one, a steam cloud almost always rising from its summit, but adverse weather kept them away. After a week of waiting, they also gave up hunting and returned to King Cove, then to the larger community at Unalaska.

At Unalaska, Jaggar was invited to join a Coast Guard crew. For the next month he and the crew sailed the full length of the Aleutian Islands reaching the westernmost one, Attu Island, just two hundred miles from Russia and the Kamchatka Peninsula. Most of the days they sailed through the usual fog and gales. On the return, they anchored off Chugul Island where they rescued two Aleutian men and a boy who had been marooned when their schooner wrecked. They also made stops at Umnak Island, where the Coast Guard set explosives and blew up a schooner that had sunk recently and was blocking a shipping lane, and Amchitka

Island, where Jaggar spend a day climbing the shore cliffs, finding only sedimentary and no volcanic rocks. And they made a special stop at Bogoslof Island, knowing that Jaggar had been there twenty years before.

They landed amid a herd of sea lions, which hustled into the sea, and myriads of sea birds, which took off suddenly, circled once and settled again on the rocks. There was again a central steaming lagoon, though only a foot or so deep, the bottom coated with a thick orange-colored film. And there were impact craters made in the sand by recent volcanic bombs, indicating recent explosions. The most recent explosion reported by passing ships had been the previous November. There was also a craggy mound of lava—as there had been in 1907—steaming vigorously. Over the next few years, crews of passing ships would record its slow rise and, in 1931, report that it disappeared after the next explosion.

———•———

Jaggar was back at Kilauea in October. He and the observatory's staff set to work converting the Ford Model-T chassis into a true amphibious vehicle.

They secured two oak beams to the frame, then, to the beams, attached a flat-bottom wooden skiff. The vehicle would be steered in water, as it was on land, by turning the front tires, which acted like rudders. To power the vehicle while in the water, Jaggar had two steel paddle wheels, each one twenty-four inches in diameter, added, driven by a chain attached to the rear power axle of the Ford chassis. The overall length was twenty-one feet. The maximum crossway measurement was five feet, four inches.

On December 21, 1927, Jaggar drove his boat-car thirty miles south along an unpaved road to the beach at Ninole to make the maiden voyage. The vehicle entered the water on its own power, then, after a wide turn in deep water, climbed back onto the beach. Jaggar would report that his vehicle "went forward and backward as required." It also rode deeper in the water than expected; the paddle wheels were almost completely immersed and seawater rose almost to the gunwales. To remediate this, Jaggar bought washboards from a local store and added them to the sides of the skiff to give it more freeboard. Another short sail was made around Ninole Bay, then the vehicle was driven back to the observatory.

The first public test came on January 17, 1928, at Hilo Bay. Hundreds of people lined the sandy shoreline, anxious to see what would happen. Some made wagers that the boat-car would never be able to climb back on shore on its own power. The more pessimistic ones said the first large sea swell would swamp the vehicle and send it to the bottom of the bay.

Jaggar made a few runs back and forth along the beach, stopping several times to make adjustments. When all was ready, he drove down to the water's edge and stopped. Isabel took out a bottle of ginger ale—this was the era of prohibition—and broke it across the bow, christening the vehicle the *Ohiki*, the Hawaiian sand crab. Then Jaggar waved to everyone and drove it toward the water.

Just as it touched the water, shouts and cheers rose from the crowd standing on the shoreline. As it got to deeper water, the paddle wheels were set in motion, moving the *Ohiki* slowly away from shore. Jaggar stopped and made a few more adjustments. Then he headed out into the bay, making a large circle and arrived back up on the beach. Afterwards, he gave rides to those who wanted to go, especially children, taking them out nearly a mile to the breakwater and bringing them back.

The big test came a month later when Jaggar began a drive-and-sail of the *Ohiki* around the island. Thurston arrived; he had financed the building of the vehicle. The two men, Isabel and two mechanics left the observatory on February 8. It took two weeks to circumnavigate the island, driving nearly two hundred miles on land and sailing more than fifty miles by sea. They camped as they went. All went well until they got to the west side of the island where the beaches are of loose sand. Here the rubber tires of the *Ohiki* dug into the loose sand or slipped and spun on mud. The remedy was to use wooden planks and metal gratings, borrowed from local ranchers, to improve the traction. Then, during a slow climb of a long road grade, the rear axle broke. It took an entire week to fix.

The *Ohiki* never proved itself to be a practical vehicle; it would never have withstood the rough sea conditions of Alaska. But it was a start. From it, Jaggar had learned much about maneuvering onto and off of beaches, about freeboard and about mechanical propulsion in water. He was ready to design and build his next boat-car when, much to his surprise, in the July 1927 issue of *Motor Boating* magazine, he learned he could buy a small one.

———•———

George Powell was a self-described "yacht machinist." He owned several patents, most for devices that stabilized yachts and small boats. He was also experienced at modifying automobile engines and installing them in various types of small marine craft. And that led to his first and only commercial product: a mobile-boat that could be used by duck hunters.

The mobile-boat that Powell invented was remarkably similar to the *Ohiki*. Both were based on a Ford Model-T chassis, the mobile-boat using a metal skiff, while Jaggar had used a wooden one. The steering of both was controlled by turning the front wheels, whether on land or in the water. Jaggar had used paddle wheels to propel the *Ohiki* in water, while Powell relied on the turning of the rear wheels by the Ford engine. Jaggar had designed the *Ohiki* to travel through ocean surf; Powell's mobile-boat could travel over swamp grass or shallow ponds. They were clearly thinking along similar lines. Jaggar contacted Powell. They agreed to design and build a larger version of their two vehicles, one that was durable enough to use in Alaska.

Jaggar was already planning to return to Alaska during the summer of 1928, his trip sponsored by the National Geographic Society. He contacted the president of the Society, Gilbert Grosvenor, who approved the building of the new vehicle. In just three months, Jaggar and Powell had it designed and built. It was the world's first durable amphibious vehicle.

It was twenty-one-feet long and five-feet, four-inches wide—the same as the *Ohiki*—the dimensions limited by what could be transported by railroad car. The hull had double walls of sheet steel that were riveted, then galvanized. Twin propellers replaced the paddle wheels. An engine taken from a Ford Model-T provided power for the paddle wheels. The shifting of a worm drive could direct engine power either to the twin propellers or to the rear wheels, which, this time, were of hard solid rubber. Steering in water was by rudder. There was an enclosed cabin with watertight compartments fore and aft and accommodations to sleep up to three people. Steel mats eight-feet long were included which could be used to cross soft beaches and could be rolled up when not in use. A power winch and an iron bar for heaving were added if the vehicle had to be levered out of soft ground. In all, the basic vehicle weighed nearly two tons and could carry another half ton of equipment and provisions.

Powell built the vehicle at his shop in Chicago. He delivered it personally to Jaggar in Seattle in April. They took it out on several tryout runs in Puget Sound, and several minor changes were made. Then Jaggar and the new vehicle—again, christened by Isabel, this time with the name *Honukai*, Hawaiian for sea turtle—headed for Alaska.

The goal of the expedition, which was sponsored by the National Geographic Society, was to explore and produce a topographic map of 2,500 square miles of the Alaskan Peninsula centered on Pavlof Bay. The existing maps, few in number, were based on Russian charts, some dating from the 18th century. As Jaggar would discover, there were obvious inaccuracies. Some showed bays that did not exist. Others indicated the entrances to small insignificant inlets that turned out to be the openings to large valleys.

Jaggar began his 1928 expedition at Squaw Valley about two hundred miles from the entrance to Pavlof Bay. Here he met the other members of the expedition. There was the topographer on loan from the United States Geological Survey, Clarence McKinley. The photographer was Richard Stewart, on his first field assignment for the National Geographic Society. In years to come, Stewart would become one of the Society's most celebrated photographers. John Gardiner and Peter Yatchmenoff—the same two men who Jaggar had traveled with the previous summer—worked as field assistants, as well as hunters and cooks.

It took a day for a scow from a local cannery to take the men and their equipment and provisions and the *Honukai* to the entrance to Pavlof Bay. All were unloaded from the deck of the scow. The men would now be on their own for the next three months.

Their first camp was in a protected cove with clamming flats nearby. Scores of big, olive-drab gull eggs could be found on rocky ledges. When boiled, they were scarcely distinguished from hen eggs in color and flavor. Caribou, seals, red foxes and bears were also abundant. After a daylong trek, Jaggar wondered "whether we could get back to the boat without stepping on a bear."

The land was mostly barren ground, except for the occasional patches of grass or belt of thick alders. As soon as they arrived, Jaggar spotted "low ridges of obvious geologic interest" close to camp that he planned to investigate. But in the distance was the main interest: a line of five volcanic cones, each one a potential Vesuvius-in-eruption.

The two largest and, therefore, dominant volcanic peaks were Pavlof and Pavlof Sister. The origin of the names is not clear. Russian explorers in the 1760s were the first to report the peaks. "Pavlof" is the Russian name for "Saint Paul."

The other three slightly smaller cones had not yet been named. And so Jaggar named them. The single one to the east of Pavlof he named Mount Dana after James Dwight Dana who had come to Kilauea in 1840 as a member of the Wilkes Expedition and who was the first geologist to see and describe the volcano. The cone immediately west of the twin Pavlof cones he named Mount Dutton in honor of Clarence Dutton who had written the first long scientific treatise about Hawaiian volcanoes, based on a summer-long trip he had made in 1882. And the fifth cone he named Mount Hague for

Arnold Hague, the man who had first taken him into the American West and to Yellowstone and who had inspired him to follow a scientific career by living on a frontier.

For the first month the weather was raw and cold. It snowed on May 24. Driving sleet fell on the evening of May 27. By morning, the ground was covered with fresh snow.

McKinley did his work, making barometric readings and taking photographs as he continued to prepare a topographic map. Gardiner and Yatchmenoff assisted him. Stewart worked alone, crossing the countryside, photographing plants and animals, the occasional waterfalls and the spectacular barren landscape. He found that the air was clearest at 9 P.M. when the sun was close to setting. Jaggar, for his part, kept close to camp, collecting specimens of flowering plants and mosses and seaweed and studying the progress McKinley was making on his map. Jaggar also collected fossils, finding that few were of animals. Most were of huge trees—cedars and firs—which meant the climate here was once very different.

When the weather was good, he ran the *Honukai* in high gear on the main beach and in low gear over the rippled low-tide flats. Only once was it stuck in soft ground. The rolled-up steel mats now proved their worth.

One day when he was alone in camp, he climbed a nearby steep cliff in order to reach a plateau of open tundra covered with violets, buttercups and other types of wild flowers. After a day of collecting, ready to return to camp, he chose a gully and started down it. At the same time, unknown to him, a bear was climbing up in the other direction.

If each had kept his course, in a second or two, they would have run into each other. But each one stopped. Jaggar slowly raised his rifle while the bear, a two-year-old Kodiak, leaned back, preparing to charge. Fortunately, both man and beast paused to take a measure of each other. Then Jaggar ever so slightly lowered his rifle. The bear, apparently seeing the action, relaxed. For several moments both were still. Then the two turned away, each one retreating in the direction he had come.

By late June the weather cleared, and Jaggar decided to attempt an ascent of Pavlof volcano. The other four joined him. They rode the *Honukai* to the base of the volcano. It took a day of climbing to reach the lower snowy slopes. On June 27, 1928, they made a dash for the summit, beginning just after midnight in the eternal twilight of an Alaska summer. The climb was straight up the steep slope. Along the way they startled three Kodiak bears, a mother and two year-old cubs that bounded away, scared by the intruders.

The five men continued to march upward. By midday they stood at the summit, the first recorded climb of this volcano.

"No pen can describe adequately the panorama," Jaggar later wrote of the view. On the north side of the summit crater was a large gash where the whole north rim of a former circular crater had fallen away. The crater itself was about a quarter mile across. A cloud of steam was lazily rising up from the crater floor. The geometry of the scene reminded him of how a man's waistcoat opens at the collar. The crater rim with the deep gash is the opening of the collar. The waistcoat is the barren lower slope. And coming out through the gash, or collar, was a high lumpy jumble of rock—a cravat—that, on closer inspection, proved to be comprised of volcanic bombs and ash of recent explosive activity, much of it covered with snow and ice.

And beyond were the other four peaks. Mount Dana also had a summit crater, much larger than the one he was seeing at Pavlof. Pavlof Sister was actually a large subsidiary cone of Pavlof, glistening with ice. It had no summit crater; instead, a lava plug capped its summit. Much farther away were the symmetrical cones of Dutton and Hague, less distinct because of their distances, though their slopes, too, were regular and smooth.

A week or so before they departed, Jaggar was standing alone on a grassy plain looking at the five peaks, a part of the world that, even today, few people have ever seen. As he watched, only the summit of Pavlof puffed steam occasionally. As he stood there, he removed his hat, awed by the silence of these five great sentinels that had withstood for many centuries Arctic gales of the Bering Sea. Some day, as he knew, each one would burst as a fiery furnace.

———•———

As stated at the beginning of the chapter, the work of a volcano observatory follows many different lines of research simultaneously, as was seen with Jaggar's investigation of the lava lake, the flowing of lava during eruptions of Mauna Loa, the recording of earthquakes, and the episodic rise and fall of the ground surface as molten lava moved within and erupted from Kilauea. And then there are the unexpected benefits. Foremost among these, during the first two decades that the Hawaiian Volcano Observatory existed, was the prediction of tsunamis.

Commonly, though mistakenly, called a "tidal wave"—it has nothing to do with the twice daily rise and fall of the sea—a tsunami is a series of giant sea waves, caused by an earthquake or a volcano explosion, that run up and inundate a coastal area, sometimes for many miles, causing wide-spread destruction and often death. One of the most deadly such disasters in recent history occurred in 2004 in the Indian Ocean when nearly 300,000 people were drowned by such a series of waves. The association of tsunamis with major earthquakes—such as the association of the 2004 tsunami with a major earthquake in Sumatra—has been known for much of history. What was not clear at the beginning of the 20th century was whether earthquakes were the primary cause of tsunamis—and whether the arrival times of giant sea waves at points far across an ocean could be predicted.

As to the cause of most tsunamis, the debate was greatly stirred after the Great Meiji Sanriku tsunami of 1896. A slight shaking was felt along the northeast coast of the main island of Honshu at 7:32 P.M. on June 15 of that year. But such slight shakings had been common for the last few months, and so the people who were living along the coast and who, as circumstances would have it, were at festivals celebrating the return of soldiers from the Sino-Japanese

War, paid no attention to the possibility of a tsunami. Thirty-five minutes after the ground shook slightly a giant wave did roll in and more than 20,000 people drowned. It remains the deadliest tsunami disaster in Japanese history, still eclipsing the 15,550 people who drowned along the same section of coastline by the tsunami produced by the strong ground shaking during the Tohoku earthquake in 2011.

But why had such a devastating tsunami been produced in 1896 by such slight ground shaking? That question continued to puzzle Japanese scientists. Geologist Bunjiro Koto at the University of Tokyo suggested that an undersea volcanic explosion had caused the ground shaking and produced the tsunami. His suggestion was inspired by the 1883 explosion of Krakatoa that produced a 100-foot-high sea wave that drowned more than 36,000 people. Others suggested that it had been a coincidence that an earthquake had been felt about half an hour before the tsunami because many other earthquakes had already occurred without producing such giant waves. In that case, so it was suggested, what had drowned so many people in 1896 was a rogue wave produced by a violent windstorm out in the Pacific Ocean. Such a wave had been produced, so Fusakichi Omori at the Imperial College in Tokyo argued—he was the man who Jaggar had visited in Japan—because an ocean had a natural frequency of oscillation that allowed seawater to slosh back and forth like a "fluid pendulum." If that was the case, then there was little hope that anyone would ever be able to predict tsunamis. But there was another possible explanation.

Akitsune Imamura, who was also at Imperial College, suggested the 1896 tsunami had been caused by an earthquake. Moreover, he suggested—and this was still a radical idea—it was not the ground shaking that had produced the tsunami, but a permanent uplift or drop of the sea floor. He also suggested that a tsunami is a special type of ocean wave, one that propagates at an incredibly fast speed across an ocean and that can barely be detected in deep water, but becomes larger and more powerful when passing through shallow water and approaching shore.

If Imamura was right, then, because of the simple fact that earthquake waves travel through the solid earth faster than waves through an ocean, it should be possible to predict, after a seafloor earthquake was detected, when a tsunami should arrive. The question was: Who might provide such a prediction?

———•———

The 1896 Great Meiji Sanriku tsunami did impact the Hawaiian Islands. It swept away the wooden wharfs at Kawaihae and Kailua on the west side of the island of Hawaii, even though both wharfs had withstood the impact of storm surges for many years. At Hilo, it entered the bay as an eight-foot-high surge of water, pulling small fishing boats from their moorings and swamping them and washing cargo ready for shipment off a wharf.

And this was not an unusual event. A dozen such waves had come into Hilo Bay in the 19th century. And each one had arrived unexpected.

The highest wave swept in on November 7, 1837. Captain James Lawrence was on his whaling ship *Admiral Cockburn* in Hilo Bay at the time of the tsunami. He saw the water retreat and "a great part of the bay was left dry." Then, to his amazement—and horror—a giant wave washed over his ship. He survived, as did his ship. Afterwards, he told his officers and crew that "he could drink no more, swear no more and chase whales no more on the Sabbath." One person decided to record the height of the wave where it came on shore by climbing a coconut tree and driving a ship's spike at the high-water mark. The spike was twenty-one feet above sea level.

All of this points to the importance of knowing in advance when such waves would arrive. Though he had no forethought that his work at Kilauea might confirm or refute Imamura's idea that tsunamis are caused by large undersea earthquakes, in hindsight, it is clear that Jaggar was ideally situated to test those ideas. He was located in the center of the Pacific Ocean, and so he would have direct knowledge when such waves passed through the Hawaiian

Islands. He also had the equipment that could record distant earthquakes—the Bosch Omori seismograph.

An earthquake sends out a wide variety of waves. The body waves—that is, the familiar P- and S-waves—cause the ground to shake back and forth several times a second. Because of the rapid oscillation, the energy of body waves dissipates quickly as they travel through the earth. In contrast, surface waves, which, as the name suggests, travel along the earth's surface, oscillate the ground at much slower rates, say, a few times a second. The energy also dissipates much slower, so that for a distant earthquake, the most prominent seismic wave is that of a surface wave. This means that, after a little experience of looking at seismic records, it is easy to determine whether an earthquake is local—has fast vibrations—or is distant. That is how Jaggar could differentiate between earthquakes that were originating within Kilauea or Mauna Loa and distant earthquakes farther out to sea that might produce tsunamis.

The first distant event came on September 7, 1918, when, according to Jaggar, "a powerful 'world-shaking' earthquake agitated the seismograph." Eight hours later, a large wave swept into Hilo Bay, tearing fishing boats from their moorings. Four years later, on November 10, 1922, another distant earthquake was "well registered" by the Bosch-Omori seismograph. And Jaggar told Finch that a tsunami might enter Hilo Bay within the next several hours. And one did, causing considerable excitement among local fishermen who were unprepared.

A pattern had been established. And so, on February 3, 1923, when a third distant earthquake was recorded, a telephone call was made to the Hilo harbormaster informing him to expect a large sea wave about noon. This time the word went out. The local fishermen took their small boats out to deep water. Families who lived along the coast went to the homes of friends. And the Hilo wharf was cleared of cargo. At thirty minutes after noon, the first wave did arrive, the water rising as high as eighteen feet. It was the world's first prediction of a tsunami.

But then came three false alarms. In each case, a distant earthquake was recorded, a warning of a possible tsunami was sent out, then no destructive sea wave materialized. The last one, on March 6, 1929, was particularly vexing. The signature of the earthquake on the Bosch-Omori seismograph was identical to the one recorded in 1923, only larger. That prompted Jaggar to send a radio message to Rear Admiral George Marvell, the Navy commander at Pearl Harbor: "Dangerous earthquake today. Likely to make tidal wave tonight about 10 o'clock. Use every precaution."

Marvell immediately cancelled a formal dinner planned that evening for Navy officers and ordered them to gather their crews and return to their ships. All coastal stations were evacuated. The warning also went out to the civilian community. Coastal areas on all the islands were evacuated on all islands. Ships and small boats either sailed for deep water or put out more mooring lines. Then people waited. By morning, no wave higher than a two-foot surge was seen anywhere on the islands. In fact, a person who was among the crowd of excitement seekers waiting for something to happen at Waikiki Beach would record that he saw nothing that night "but an unusually calm surf."

There was now public criticism of Jaggar and his tsunami predictions. One newspaper editorial suggested that whatever might be saved by an occasional accurate warning was offset by the mayhem caused by a string of false alarms. But Admiral Marvell saw it differently. He told Jaggar to continue to inform him personally whenever there was a chance of a huge wave. The next warning came on March 2, 1933.

Again, probably because they were acquainted with Jaggar, especially through his remarkable amphibious car, the local fishermen took their small boats out to deep water. Cargo was removed from the Hilo wharf. And Admiral Marvell gave orders to secure the Navy's ships. But that is as far as preparations went.

Jaggar predicted the wave would enter Hilo Bay at 3:30 P.M. A three-foot-high wave did come into the bay at 3:36 P.M. It was too small to do any damage. But, elsewhere, the story was far different.

On the west side of the island of Hawaii the ocean level dropped until it was eight feet below normal low tide. Then, in a matter of minutes, it rose as a surge ten feet high. Stone walls were pulled down. Boats capsized. Houses were moved and furniture ruined. Automobiles and trucks were flooded so that engines were damaged by sand. Cargo was washed off wharves. And it was the same on the other islands. As a bit of irony, the boathouse in west Hawaii where Jaggar sometimes stored the *Honukai* was shoved six feet off its foundation.

———•———

Jaggar summarized the event this way: "The seaquake wave of March 2, 1933, successfully demonstrated the methods of scientific forecasting that had been under study at the Hawaiian Volcano Observatory." The fact that the warning was largely unheeded was probably due to the false alarms of the previous years. That fact that he was able to provide any successful predictions is amazing in view of what he actually knew.

Communication was certainly not what it is today. Except for 1933, reports that a large earthquake had occurred somewhere in the Pacific, specifying the exact time and place, did not reach the Hawaiian Islands until days after the event. In 1933, a report was received by a news broadcast four hours after the event, describing an earthquake disaster in Japan that had killed more than 1,500 people and had caused a large wave to swamp the coast. Before that information arrived, Jaggar admitted he did not know where the earthquake had originated, speculating that it could be Japan, Kamchatka or the Aleutians.

And then there was the matter of the false alarms. The seismic event recorded in 1929 was remarkably similar to the one in 1923 that did produce a tsunami because both events occurred in a similar part of Alaska. But the 1923 quake was under the sea, while the one in 1929 was on land, which explains why the later event did not produce a tsunami.

And there was the question whether an approaching wave was going to be a few feet high or more than ten feet high. That is still a difficult question to answer today, even though there are now many more seismographs operating, the contours of the Pacific floor have been mapped, and instruments specifically designed to record passing waves are in continuous operation on floating buoys scattered across the Pacific Ocean.

The 1933 disaster should have led to a tsunami-warning system, at least for the Hawaiian Islands, but it did not. Instead, it took another disaster—and many deaths—before one was realized.

Jaggar was retired and living in Honolulu when the next devastating wave struck the islands. It was April 1, 1946, and a huge earthquake had shaken southern Alaska. The first seismic wave of the quake reached Hawaii and was recorded on the Bosch-Omori seismograph at 2:06 A.M. It was more than five hours before someone arrived at the volcano observatory to begin the daily work, which included examining the seismic record. By then, the tsunami produced by the Alaskan earthquake had already swept across the Pacific Ocean and slammed against the Hawaiian Islands.

The wave height reached thirty-six feet on Oahu and thirty-three feet on Maui. In Hilo Bay, it rose to an astonishing fifty-five feet. A total of 159 people drowned. Buildings were destroyed. Cars and trucks were swept away. Ships were heavily damaged.

Damage also occurred elsewhere in the Pacific. Destructive waves ran up beaches at Coos Bay, Oregon, and Half Moon Bay in California. Similar waves swept onto the Marquesas Islands of French Polynesia and along the coast of Chile.

Because damage was so widespread, in 1949 the world's first tsunami warning center was established on Oahu. It traces its origin to the telephone call Jaggar made in 1923 to the Hilo harbormaster warning him of the possibility of a devastating wave.

The 1920s was a period of accomplishment for Jaggar. It began with revealing the episodic rise and fall of the surface of volcano as molten rock was supplied to, then removed from the lava lake.

It continued with following the progress of the 1926 eruption of Mauna Loa. Then an unexpected transfer of the volcano observatory to the United States Geological Survey opened to door to a greatly expanded program of volcano research. And that led him back to Alaska for two summers. He built the world's first two practical amphibious vehicles. He hired more people to work at Kilauea. He purchased more scientific equipment, especially seismographs, some of Omori's design and some of his own. And better topographic maps were being made of the island of Hawaii.

And, then, it all came to a halt.

The economic depression, which, in part, was precipitated by the crash of the New York stock market in October 1929, was felt in full force by Jaggar four years later. To deal with the worsening economic situation, in 1933, the newly elected president, Franklin Roosevelt, started many new government programs. He also curtailed many existing ones. One of those that was cut severely was the volcano program. In 1933, the budget for volcano research was reduced by half. In 1934, it was eliminated entirely.

The work in Alaska ended. The volcano observatory at Lassen Peak in California was closed and Finch joined the growing number of unemployed. The volcano observatory in Hawaii was also to be closed. It then had six employees, including Jaggar. The other five did lose their jobs. Jaggar would also have lost his job and the observatory would have closed if circumstances did not intervene.

The intervention came from the superintendent of Hawaii National Park, a former North Carolinian, Edward Wingate. In 1921 Wingate had scored the highest on a civil service exam for civil engineers. That allowed him to choose any position then available. He looked at the list and chose what he thought would be the most exciting one—working with Jaggar at Kilauea. In 1931 he took the position of park superintendent. Three years later, when it looked like the Hawaiian Volcano Observatory would be closed, Wingate contacted his father-in-law, Shelby Singleton, a prominent Chicago lawyer, whose childhood friend was Harold Ickes, now the Secretary of Interior. It was through this fortunate trail of political connections that, on July 1, 1935, the Hawaiian Volcano Observatory became part of the National Park Service and Jaggar was given the new position as park volcanologist.

Admittedly, the volcanic activity in Hawaii was much less during the 1930s than it had been during the first twenty years the observatory had existed. Jaggar termed it simply: Kilauea "went to sleep."

He retired on July 31, 1940. On that day, as he recorded it, "the whole Park gang" came to his office with drinks and ice for a farewell party. The next day he went to Hilo and sailed the *Honukai* around Hilo Bay. The remainder of his life would probably have been one of writing scholarly papers. But, then, the Second World War began.

CHAPTER SIXTEEN

A FORGOTTEN LEGACY

The Jaggars moved to Honolulu on Oahu, as the University of Hawaii had appointed him to a professorship that he would use to write about his years studying volcanoes. Leaving their house on the edge of the lava lake, the Jaggars built a new house near the university. The ground floor had a single large room. To one side were a kitchen and a small dining area. A single bedroom was upstairs. To one wall of the bedroom Jaggar nailed a ladder. At the top of the ladder was a trapdoor that led onto a flat roof where he mounted a telescope. They spent many of their evenings sitting on the roof looking through the telescope at the moon or simply gazing at the night sky.

They left their house at Kilauea under the care of their housekeeper, Shizuka Yasunaka, who had worked for them since their marriage. The general upkeep of the house was the responsibility of Shizuka's husband, Hideichi, who had worked for many years at the volcano observatory. His many tasks were described by Jaggar "as janitor, messenger, automobile mechanic, and outdoor man, he assists in operation of seismographs, and in recording the meteorological data, he sends weekly reports to the Weather Bureau and is invaluable as general caretaker of the grounds, the plumbing, carpentering repairs, and painting."

On November 22, 1941, a strong earthquake shook the island of Hawaii. Afterwards, Hideichi went to inspect the Jaggar house and its furnishing. The next day he mailed a short note, informing the Jaggars that the shaking had "caused excitement, but no damage." On December 3, Hideichi wrote again, saying that the $30 the Jaggars had sent him for maintaining their house at Kilauea was $5 more than he was owed and that he was returning the difference by mail.

Four days later, on Sunday morning, December 7, 1941, the Jaggars were attending mass at St. Andrew's Cathedral in Honolulu when Pearl Harbor was attacked. They watched as Japanese planes flew low overhead and heard the explosions of bombs and saw black dense clouds of burning oil rising from the harbor. The next few nights they and many of their neighbors slept in the basement of the library at the university for protection from possible future attacks. Rumors were rampant. All news had been suppressed. There were no commercial radio broadcasts.

On the afternoon of December 7, martial law was declared in the Hawaiian Islands. The movement of civilians was restricted. Those who were Japanese nationals had additional restrictions that prohibited them from changing residences or occupations or "otherwise travel or move from place to place." Those additional restrictions applied to Hideichi and Shizuka Yasunaka: He had been born in Japan; she had been born on the island of Hawaii before annexation, and so United States government regarded her as a Japanese national.

The additional restriction meant they could not go to stores and buy goods. And so they sought help from Thomas Jaggar, writing to him, asking if he could send a variety of garden seeds. On February 19, 1942, he wrote back saying he had purchased five pounds of cabbage seeds, the only type of seed he could find, and was sending it to them immediately.

The same day President Franklin Roosevelt signed Executive Order 9066 that authorized the evacuation of all persons who were deemed a threat to national security from the West Coast to relation camps inland. The same day, an order was sent from Washington, D.C., to the military governor of the Hawaiian Islands, Army General Delos Emmons, who was told to gather all persons of Japanese ancestry who lived in the islands and intern them on the small island of Molokai for the duration of the war.

Emmons balked at the order. There were more than 150,000 such people living in the islands, and, as Emmons knew, they were essential to the economy of the islands, and, hence, to the war effort. More than half worked on sugar plantations. Nearly all the truck drivers and carpenters were of Japanese ancestry. He informed people in Washington, D.C., who modified the order. In their estimation, about ten percent of the Japanese-American population in the islands represented a security risk, and so those people were to be sent to the United States mainland for interment, beginning, as the new order read, "with the most dangerous." Again, Emmons balked, delaying his response for three months. Finally, he wired the War Department that, according to his sources, about one percent of the Japanese-American population might be a security risk. The War Department agreed. And so a quota was set. Emmons would have to send 1,500 people to these internment camps.

———•———

Francis Kaneaki Yasunaka, the youngest of the four Yasunaka children, would always remember the day. He was nine years old. He had come home one afternoon and, to his surprise, his mother was not home. She was always home. He went to nearby houses to search for her. He found her crying—as she would throughout the night—and continued to say: "They took dad away. They came and took dad away."

That morning, Wednesday, July 15, 1942, two men arrived unexpectedly at the Yasunaka house. One man was from the Office of Navy Intelligence and the other from the Federal Bureau of Investigation. The Navy man took Hideichi Yasunaka to a nearby military camp and put him under guard. The other man searched the house, finding a photograph of the Emperor of Japan and a medal won by Hideichi's father during the 1905 Russo-Japanese War.

A military tribunal of three officers was held that afternoon at the camp. They were shown the items taken from the Yasunaka's house. They heard testimony—from whom was never disclosed—that Yasunaka had been seen socializing with officers of the Japanese Navy before the war.

That was true. Since the 1920s, ships of the Imperial Japanese Navy did visit the Hawaiian Islands, and, when they did, people were encouraged to welcome the ships' officers and crews. Stores and hotels were decorated, and lanterns strung along streets. During one of those visits, in 1925, a squadron of ships called at Hilo. More than three thousand Japanese officers and sailors came ashore, many of them boarded in private homes, the residents encouraged by the United States Department of State to do so. And they took tours of Kilauea. Since Yasunaka was the only one of the observatory staff who spoke Japanese, he led the tours. As a show of appreciation, that night officers of the Japanese Navy hosted at a dinner at the Volcano House in Yasunaka's honor. It was this association—he had given several tours to visiting Japanese Navy officers and sailors—and having in his possession a photograph of the Emperor and a Japanese military medal—that the three officers of the tribunal pronounced Yasunaka a security risk. He was sent to the Navy yard at Sand Island on Oahu to await transfer to an internment camp. He contacted Jaggar for help.

Jaggar knew the military governor, General Emmons. They had worked together planning the bombing of Mauna Loa in 1935. Jaggar wrote to the General, on behalf of Yasunaka and of another former employee of the observatory, Asao Okuda, who was also being held at Sand Island. Okuda had been a mechanic at the observatory and, at the time of his arrest, was living on Oahu, working for Jaggar at the university.

"Hideichi Yasunaka has served me since 1924," Jaggar wrote at the beginning of a letter to General Emmons. "He has exhibited perfect loyalty to me and the service of the Hawaiian Volcano Observatory for many years." Then, to reassure Emmons, Jaggar added, "I am convinced that this man is not dangerous. Both Mrs. Jaggar and I are willing to vouch for his loyalty to the United States." A similar letter was sent for Okuda.

Emmons authorized Okuda's release provided either Thomas or Isabel Jaggar accompanied him wherever he went. They built a small house on their property near the university where Okuda and his wife lived for the remainder of the war. The case of Yasunaka did not have such a favorable outcome.

"The information contained in your letter has been weighed," wrote Emmons. "The conclusion reached is that the original order was fully warranted. The request for his release is, therefore, denied."

Shizuka Yasunaka now had a difficult decision to make: whether to remain in the islands with their children, or have them and her accompany her husband to an internment camp. She chose the latter.

She packed seven boxes, all that she was allowed to bring. Five she filled with clothing. Another contained her husband's carpentry tools. In the last she packed her sewing machine, not knowing where they would live or under what conditions.

On March 2, 1943, the Yasunaka family was in the final group of Japanese internees to leave the Hawaiian Islands. Two weeks later, they reached a desolate spot near Topaz, Utah. Here they entered a camp, a square mile of tarpapered wooden barracks surrounded by barbed wire and armed guards. Each barrack had a similar division of rooms: one room to a family. Each room had a potbelly stove and army cots. The sole source of light was a bare bulb that hung overhead.

After the end of the war, the Yasunaka family was released and returned to the Hawaiian Islands. The oldest son, Gary, then enlisted in the United States Army and was sent immediately to work as an interpreter in Japan. From that experience, he decided to make the Army his career. On April 25, 1951, he was part of an artillery group that was overrun during a Chinese offensive in Korea. He is listed today as "missing in action, presumed dead."

Eventually Hideichi Yasunaka killed himself over the grief of his lost son. Shizuka then moved the family to California where I met her as I was doing research for this book. Because the family's tragic story could be traced back to Hideichi's association with the Hawaiian Volcano Observatory, I was concerned how she would receive me.

All apprehension disappeared as soon as I entered her house, because the first thing I saw, hanging prominently on a wall, was a portrait of Thomas Jaggar.

During the short time we talked, I asked her what was the daily routine when she worked for the Jaggars. She said that her family lived in a small house between the Jaggars' house and the observatory. She waited each morning until both the Jaggars had left, then she went over the cleaned the house and collected the laundry. She returned in the afternoon to cook dinner. Pork chops was Thomas Jaggar's favorite meal.

One last significant point: As we talked, I noticed that Shizuka, who was then 102 years old, still referred to Thomas and Isabel Jaggar as "Papa" and "Mama."

Ten days after the attack on Pearl Harbor, Major James Snedeker of the United States Marine Corps came to see Jaggar. They had met in October 1940 when Snedeker was at Hilo Bay watching a demonstration of the *Honukai*. At the time, the United States military had no such vehicles, while Jaggar had already designed and used two. Snedeker was impressed by the *Honukai* and had it shipped to the

Marine base at Kanehoe on Oahu where engineers disassembled it and used it as the basis for a new type of military vehicle, the DUKW (pronounced "duck"), which would transport troops and goods over land and water. More than two thousand such vehicles were produced during the Second World War, playing a critical part in amphibious landings in the Pacific and on D-Day in Normandy in France.[*] For this contribution, after the war, Jaggar received the Franklin Burr Award for scientific innovation from the National Geographic Society. But, now, as the war in the Pacific was beginning, Snedeker had a special request from those who would be planning the war.

Jaggar had traveled to places that would soon be major battle-grounds. He had been to Japan four times, the Aleutian Islands three times and the South Pacific twice. Based on his travels—and his scientific knowledge—Snedeker asked that Jaggar write a series of essays for the United States Navy that would be useful in the conduct of the war. In all, Jaggar would produce 2,205 essays, most one or two pages in length and many illustrated by his drawings or photographs. They included such titles as "Aleutian Shorelines," "Disaster Expeditions in Japan" and "Steamblast Disasters." After the war, Jaggar assembled a small part of his essays into a book, *Volcanoes Declare War,* which he dedicated "to helpmeet and camp-mate Isabel Jaggar whose horse crushed her against a tree . . . Whose gloves fell into a red hot crack and burned up . . . Who slept in a lava tunnel beside the immortal remains of a desiccated billy goat . . . And loved it all."

———•———

With his days of adventure, travel and field work behind him, Jaggar's lasting legacy, though now largely forgotten in scientific circles, extends far beyond the writing of essays for the Navy. His work is the foundation of almost every aspect of volcano research today.

***

[*] DUKW is an acronym that means: D=designed in 1942, U=utility vehicle, K=all wheel drive, W=dual rear axles

The first volcanic gases collected at an eruptive fissure were made at Kilauea in 1912 by Shepherd, Day, Dodge and Lancaster. Jaggar continued to collect gas samples—and improve on the technique. Of particular note are the pristine samples he collected, essentially working alone, at Kilauea in 1917 and at Mauna Loa in 1919. For Mauna Loa, his are still the only gas samples collected during an eruption of that volcano.

The sampling of volcanic gases is now routine and, as is much of the work at active volcanoes, has been automated by the use of continuous sensors. These measurements show the variability of volcanic gas emissions and continue to reconfirm that, after water vapor, carbon dioxide is the major gas emitted by volcanoes. (The measurements also show that, during the last century, the amount of carbon dioxide emitted by volcanoes, even during major eruptions, is greatly dwarfed by that emitted by cars or by industry.)

The recording of earthquakes continues to be the key to predicting volcanic eruptions. For centuries people had fled when the ground started shaking near a volcano. And they returned when the shaking stopped, whether the volcano had erupted or not, not knowing whether events too small to be felt were continuing. Starting with the 1914 eruption of Mauna Loa, with Jaggar and Wood standing in the Whitney Vault, watching the subtle vibrations recorded by the Bosch-Omori seismograph, it is clear that such activity can be recorded reliably. And, today, it is routine to follow the migration of earthquakes through a volcano as magma pushes its way toward the surface and erupts.

It took Jaggar decades to convince others that the Bosch-Omori seismograph was also recording a slight tilting of the ground and that such tilt changes were related to magma movement within Kilauea. Today the movement of magma is recorded with precision land surveying equipment and by space-based systems. The Global Positioning System is being used to record movements as small as 1/20 of an inch (about 1 millimeter) at dozens of active volcanoes. Satellites with radar equipment can reveal the rise or fall of the ground surface over a broad area—often the only way to determine if the surface of a remote volcano is moving—and whether the volcano is building toward an eruption.

———•———

As is well known, at the end of the Second World War, the United States and the Soviet Union became locked in a confrontation that came to be known as the Cold War. At the time, some military strategists thought war between the two former allies was inevitable. And so, at least in Washington, D.C., plans were made, priorities were shifted and departments reorganized. One of the small changes came as a request from the War Department to the Geological Survey to undertake a program of volcano investigations in Alaska. It was duly done. As part of the program, on December 28, 1947, the Hawaiian Volcano Observatory was transferred from the National Park Service to the Geological Survey.

Jaggar was pleased by the decision—and by the prospect of a renewed federal program of volcano research. Moreover, he was doubly pleased because the man who would run the program was Howard Powers. Powers had begun his career as a geologist by working as Jaggar's assistant after Finch moved to California to start a second observatory at Lassen. Powers had been one of the casualties during the cutbacks in the 1930s when his job was eliminated. But he returned immediately after the war. And, now that he was in charge of the new program, it seemed inevitable that Powers would fully integrate the work of the Hawaiian Volcano Observatory with the larger goals of working in Alaska.

Powers arrived at Kilauea in May 1948 to discuss the matter with Jaggar and with the three members of the observatory staff. Finch was still the director. Working with him was a geologist and a machinist.

Powers probably began by quoting a memo by the Director of the Geological Survey that stated that the new volcano program would be "centered on the observation of active volcanism near military bases" in Alaska. The work itself would consist primarily of gathering geologic information that could be of military use, for example, the nature of beaches or the frequency and violence of eruptions. From that, the military would know the most suitable places to locate harbors and airfields.

But what of the future of volcano work in Hawaii? To that question Powers had a disappointing answer.

Though the Hawaiian Volcano Observatory had recently been transferred back to the Geological Survey, its work did not fit into the goals of the new program. Hawaiian volcanoes had mild eruptions, while those of Alaska exploded, and so what was learned in Hawaii could not be applied to Alaska. Moreover, Hawaiian volcanoes had not been very active in recent years. Kilauea had not erupted since 1934. And the two most recent eruptions of Mauna Loa had been brief, the one in 1942 lasting two weeks and the one in 1943 lasting only three days.

And so, Powers continued, in order to direct all efforts toward Alaska, Finch would be retired, reducing the volcano staff in Hawaii to two men. And those two men would work for another year or so on a part-time basis and at half salaries. Their purpose would be to close the observatory.

In the aftermath, Jaggar wrote to friend, telling how established science had betrayed him. But there was nothing he could do. These were the closing years of his life. And he would have to watch how, with a wave of a bureaucratic hand, his forty years of work at Kilauea would come to an end.

———•———

On December 31, 1949, his sister, Anna Louise Jaggar, died. She had spent her working career at the botanical garden at Harvard University, achieving a small measure of notoriety as a cataloguer of plants. When she retired, she returned to the family lodge in Nova Scotia, the one their father had purchased in 1886. Throughout her long life, her brother had sent her $50 a month to supplement her meager wages.

As her only surviving relative, Thomas Jaggar made the long trip to Nova Scotia to settle legal matters and to gather family heirlooms. It was a trip that would be remembered for the four fires.

To get to the Jaggar lodge, one had to travel from Maine across the Bay of Fundy to the small community of Digby. Then it was a short train ride to Smith's Cove where the Jaggar lodge was located. Having arrived in Digby, Jaggar was waiting at the station for the next train when a man approached him. He asked if the visitor was Professor Jaggar from Hawaii. Jaggar said he was. The man said he had bad news. The previous night the Jaggar lodge had burned to the ground. Everything was lost in the flames. That was the first fire.

Knowing this would be his last trip away from the Hawaiian Islands—Jaggar was now seventy-nine years old—he decided he would spend a week with each of his children. It would be the first time he had spent any extended amount of time with either of them since his divorce from Helen thirty-five years before.

He traveled by train through Canada to Vancouver, British Columbia, then southward to Seattle, Washington. From there, he took a ferry across Puget Sound to the rural town of Poulsbo where his son lived. When he arrived, his daughter-in-law told him that his son was next door helping a neighbor bring down an old barn. When Jaggar arrived, he saw the barn was on fire and that a man was collapsed outside. It was his son, who told Jaggar that he did not know what happened. He was inside the barn when it caught fire and he barely escaped. Fire number two.

Jaggar next goes to the San Francisco area to see his daughter. But, as she told me: "It was the darnedest thing. As soon as Dad arrived, my skin broke out in a red burning rash. It must have been something I ate. And I wasn't able to spend any time with him." That was fire number three.

Jaggar now meets Isabel and they sail back to the Hawaiian Islands. On the last night at sea they are dining with the captain when he receives a cable. Mauna Loa is erupting and lava is pouring into the sea. The captain then announces that he will be diverting the ship so that everyone can see the display.

The next evening, as Jaggar recalled the scene, the ship is four miles off the Kona coast. There are two zigzag streaks of orange coursing down the mountain, each one looking "like hot coals extending far up the mountainside under the clouds." Occasionally, there is a bright flare where a tree has burst into flame. "Visible motion there was none," he recorded, "as we were too far away to see detailed motion." This was fire number four.

As he and Isabel watched the display, he confided to her that, after forty years of living at Kilauea, fire now followed him wherever he goes. She told him that the name people in the islands have given him—*Malama-o-Pele*, "the torch of Pele"—is an appropriate one. He replied, "I am the weaker one."

———•———

Jaggar's health began to deteriorate after his return. There was the occasional household accident or an illness that sent him to the hospital. Some required extended stays. His last hospital stay was in December 1952. His doctor released him a few days before Christmas and sent him home.

He had written an autobiography and had shown it to publishers on his last trip to the mainland. None chose to publish it. As a last hope, he submitted it to University of Hawaii Press, sure it would publish his memoirs. It did not.

A rejection letter from University of Hawaii Press arrived on January 16, 1953. Thomas Jaggar died early the next morning in his sleep—forty-one years to the day from when he had arrived at Kilauea to start the volcano observatory.

———•———

Kualono, the home of Thomas and Isabel Jaggar at Kilauea, was their place of privacy. It was a place where they could get away from the crowd of visitors at the Volcano House and still see the observatory and Kilauea and Mauna Loa. It was here on the verandah that he contemplated many philosophical questions. One was: What is love? And he provided an answer.

"Have you ever sat after a day of hard work beside someone you loved," he wrote, "with her hand in yours thrilling, and gentle sweet night airs blowing by, fragrant with flowers and faintly musical with the leaf whirr." And you ask her: Do you love me? And she answers, "Yes."

During his last years, he and Isabel talked of many things, as couples do. They even discussed their eventual deaths.

Neither wanted to be placed in the ground with tombstones over their heads. Both said they wished to be cremated and their ashes scattered in a special place. For Isabel Jaggar, it was the small garden she had kept near Kualono. For him, it was a different place.

It is often mistakenly said that Isabel Jaggar scattered her husband's ashes in Halema'uma'u. That is not true. There was a place dearer to him.

I have seen a letter she wrote to a friend in which she tells where she placed her husband's ashes. At the beginning of the letter, Isabel wrote: "This is a private matter." The underlining is hers.

As the writer of this book, I cannot violate a desire to have a resting place remain private. But I can say that the mortal remains of Professor Thomas Augustus Jaggar, Jr., are somewhere at Kilauea. And that is where he rightly belongs.

POSTSCRIPT

I n June 1952, after almost twenty years of quiescence, seven months before the death of Thomas Jaggar, lava again erupted from Kilauea and a lava lake formed within Halema'uma'u. The lava lake was active for 136 days.

The next eruption came two years later and continued for three days. Afterwards, the Bosch-Omori seismograph in the Whitney Vault indicated a slow swelling of the volcano that continued until early 1955 when the summit started to subside and lava erupted east of the summit at Kapoho, the same general area where earthquakes had occurred and dramatic ground cracks had formed in 1924. The 1955 eruption lasted eighty-eight days. It was the most voluminous outpouring of lava from Kilauea in more than a century. It was this eruption that prompted the United States Geological Survey to resume employing full-time workers at the Hawaiian Volcano Observatory.

Today the observatory is housed in a large building located near the highest point of the caldera rim. The staff numbers about two dozen people whose specialties range from geology to geodesy, from geochemistry to geophysics. Within the building are modern laboratories where the chemistry of volcanic gases or of lava samples can be analyzed. Computers are connected through high-speed data lines to hundreds of automatic field instruments. Any change in gas emission or seismic activity or any slight movement of the ground is instantly detected, recorded and analyzed.

And next to the observatory is a museum dedicated to Thomas Jaggar.

———•———

I made my first trip to the Hawaiian Islands in 1971. On my last night in the islands, I was in Hilo and heard on the radio that Kilauea was erupting. I raced to the summit. Thousands of people had already gathered. The eruption was two miles away. I could see a distant red glow. I stood there for a few hours. My view of the eruption ended late that night in a heavy rain.

Three years later I returned to Kilauea and spent a summer hiking trails. An eruption had occurred two weeks before I arrived and another occurred soon after I left. Years later, by a fortunate set of circumstances that are too complicated to relate here, I worked as a volunteer, then as a staff member at the Hawaiian Volcano Observatory. That was when I was introduced to the work of Thomas Jaggar. Many eruptions have followed.

Each one has been memorable. On one occasion, I was living in a house with a bedroom on the upper floor. A row of six windows lined one wall. One night, shortly after midnight, I felt the bed shake. I looked out the windows. Through the leftmost one, a few miles away, I could see a fountain of lava of Kilauea. Through the rightmost window, at a much greater distance, was the broad profile of Mauna Loa silhouetted by a bright apricot glow. It was the first simultaneous eruption of the two volcanoes in sixty-five years.

By great fortune, a lava lake returned to Halema'uma'u several years ago. I made the trip to see the glow, going at night, sometimes with a friend, often alone, standing close to the place where the Jaggar house once stood or at one of the other places of solitude along the caldera rim. Even though the crater is more than a mile away, if the air is still and the clouds are low, it is often possible to hear molten lava sloshing inside the crater.

It is at such times that I am reminded of what I have learned at Kilauea. Long after the memory of me or of the reader of this book has long been forgotten, in fact, long after the words in this book have faded beyond all recognition and the pages have turned to dust, Kilauea will still be erupting.

Because it is the volcano that is eternal.

THE END

ACKNOWLEDGMENTS

I met Thomas Jaggar's daughter, Eliza Bowne "Sallie" Jaggar, three times. During our first two meetings she was polite and cordial, but not very helpful. She kept reminding me that she had been a baby when her parents separated and divorced, and so knew little about her father. Our third meeting was considerably different.

As on the previous occasions, I telephoned the day before we were to meet to make sure I was expected. She said I was. Then, when we did meet a third time, she told me this story: After she hung up the telephone she went into her bathroom and looked in the mirror. There, over her shoulder, she saw the face of her dead brother. He told her that it was time to come with him. She answered no. She said that someone was coming to see her the next day and that she had decided to tell everything about their parents. When I did arrive the next day, Sallie did exactly that. It was like the floodgates of a dam had been opened. For almost an hour, she told me how her parents had met. She recounted their courtship and difficult marriage and how it had ended after a torrential rain when they were at the Hilo Hotel. She also revealed that, many years later, her mother had written a long memoir, never published, that was to be read only by her daughter. She then handed me several excerpts she had transcribed and said that was all she could give me. I am indebted to Sallie Jaggar for trusting me with the personal story of her parents.

I am also indebted to Thomas Jaggar himself. He left an enormous amount of written material, much of it in the form of field notes and scientific writings, but also several sermons, a collection of private letters and a set of small notebooks in which he jotted down daily happenings. In the top margin of one of the pages he has reminded himself: "Write 3000 words a day." It is a goal he tried to keep throughout his life, even during the last days of his life.

Though I first encountered the work of Thomas Jaggar in 1976, I did not begin this book in earnest until twenty years later. And then it was almost another twenty years to see it to an end. Through the second twenty years, only one person showed unwavering support for the book and was always optimistic that it would be published. Tom Peek has guided me through the many pitfalls of learning to tell stories. He also patiently read through a tortuous early draft of this book. His comments have improved it greatly. He continues to be an inspiration to me and to many others, encouraging us to write personal stories because that is how we understand who we are.

Crucial to the contents of this book were the interviews conducted with people who either knew Thomas Jaggar or who were at Kilauea volcano during the early years of the volcano observatory. Northup Castle served as a surrogate son to Jaggar. He told me of riding with Jaggar in a Model-T to the lava lake. Barbara Fitzgibbons remembered how relieved she was when Thomas and Isabel Jaggar drove up to her family's house in 1929 after a series of earthquakes shook the Kona side of the island. She said she felt better as soon as Professor Jaggar arrived. Akira Yamamoto's father was the headwaiter at the Volcano House restaurant and his mother worked in the laundry and, sometimes, in the kitchen.

Akira remembered sitting in the back of the dining room and watching Jaggar give lectures to dining patrons. Afterwards, as Akira recalled, the former Harvard professor circulated among the diners and performed magic tricks. Jaggar chose Alfred Tai On Au from a shop class of high school boys to work at the observatory. Alfred was one of the people who built and modified the two amphibious vehicles.

A special mention must be made of Shizuka Yasunaka who worked as the Jaggars' housekeeper for more than twenty years and whose story is much more poignant than I could include in the main text. She was born in 1897 in a plantation town north of Hilo. She married at age fourteen and, after giving birth to a son, she and her husband and newborn son moved to Japan where they lived in Hiroshima for two years. When they returned, she began working for the Jaggars. She gave birth to a second son. A few years later, her husband went insane and a paternal uncle took custody of the two boys. She then married Hideichi Yasunaka and, years later in order to preserve their marriage, she went with him to internment camps in Utah and in California where they spent more than three years behind barbed wire. Through it all, she never lost faith in humankind. Meeting her was one of the high points of researching this book.

Taeko Jane Takahashi served for many years as the librarian at the Hawaiian Volcano Observatory. I relied on Jane repeatedly to direct me to written accounts of early visitors to Kilauea. Ben Gaddis has taken on the task of gathering and annotating and making accessible the thousands of photographs taken of Hawaiian volcanoes and of the people who have worked here. Ben selected several of the photographs used in this book. He is also the person who relocated the Jaggar daybooks.

James Cartwright helped me navigate the archives at the University of Hawaii. Cynthia Murphy at the Connecticut Valley History Museum sent me a stack of newspaper and magazine clippings about Frank Perret. John Fournelle unselfishly provided information about Ruy Finch and documents from the national archives about Hawaii Volcanoes National Park. Katharine Cashman provided a map of the historical lava flows on the island of Hawaii.

Darcy Bevens' transcription of the *Volcano House Register* was an important aid in understanding the history of Kilauea volcano and of the Hawaiian Volcano Observatory. How she managed such a mammoth undertaking was nothing short of a miracle. Peter Charlot's *A Scientific Missionary*, a one-man play about Thomas Jaggar, gave me a unique perspective into the life of this remarkable and unusual man. Over the years Peter and I have had many discussions about Thomas Jaggar.

A special thank you is owed to Ardis Morrow who had the foresight to save Jaggar family documents and photographs. Bruce Blevins has published Jaggar's diaries and photographs for the 1893 and 1897 expeditions to the Absaroka Mountains.

I thank Liek Pardyanto, Johannes Matahelumual, Harun Said and Deddy Mulyadi for guiding me through the challenges at Galunggung volcano. In Italy, it was friendships with Roberto Scandone, Roberta Scarpa and Paolo Gasparini that helped me navigate the confusing world of Italian science.

Robert Tilling first opened the doors of the Hawaiian Volcano Observatory to me. Robert Decker saw and fostered more possibilities in me than I thought existed. Both have served as Scientist-in-Charge at the observatory. Numerous discussions with Frank Trusdell helped me hone my understanding of Kilauea and Mauna Loa. My general understanding of volcanoes has also benefitted from years of friendships with Daniel Dzurisin and Thomas Casadevall. Thomas English provided personal insights into the recent history of Kilauea and to living on the island of Hawaii. Kathy English of the Hawaii Natural History Association allowed me access to the Association's archives, which included a tape-recorded interview made by Thomas and Isabel Jaggar.

As this book was being researched and written, many people listened to me as I struggled to assemble the story of Thomas Jaggar. James Kauahikaua, Susan Dieterich and James Dieterich are among those who I relied on most—and who showed the most patience.

Keola Awong helped me understanding the meaning of Hawaiian place names. Mary Siders and Wilfred Tanigawa commented on an early version of the manuscript.

Unless otherwise noted, photographs are from the Jaggar family collection. Additional photographs were published with permission provided by Tracy Laqua of Hawaii Volcanoes National Park, Barbara Dunn of the Hawaiian Historical Society and Tia Reber of Bishop Museum, and, by the University of Hawaii, archives and papers of T.A. Jaggar.

A special thanks is owed to my agent Laura Wood who, early on, realized the importance of telling the story of Thomas Jaggar. She improved an early draft of this book. I am grateful to my editor Jessica Case for her patience when I missed deadlines and for having the foresight to publish this book. Her numerous comments on an early draft greatly improved this book.

The story of the four fires told at the end of the book is written as told to me by Sallie Jaggar.

I close with one final story about Thomas Jaggar. The first time I met his daughter I asked her: What is the earliest memory that you have of your father? This is what she said.

It was 1915 and she was four years old. Her mother had called a taxi and the three of them—her mother, Sallie and her brother, Kline, who was then ten years old—were driven to the Cliff House, which, in those days, was a stately hotel that stood along the Pacific Coast south of the Golden Gate near San Francisco. When they arrived at the hotel, her mother told her children that their father was inside waiting for them. (Helen Jaggar refused to go in.)

As Sallie remembered it, once she was inside the hotel, they were greeted by a man who said he was her father. He took them to the hotel dining room where they had lunch. Sallie could not recall anything of the meal, but she did remember that, before they were served, her father went around the room and gathered all the toothpicks from the tables. He then assembled them into a tight ball. He lit a match and burned through one of the toothpicks. That set off a minor burst as toothpicks were sent flying around the room.

After lunch, the three went for a walk along the boardwalk. A man was offering to cut full-body silhouettes from black paper. Thomas Jaggar had a silhouette made of himself and of each of his children.

Years pass. In 1987 the Hawaiian Volcano Observatory had a celebration to mark its seventy-fifth anniversary. Sallie Jaggar attended. As part of the celebration, some of Jaggar's personal items are on display. These include the three silhouettes. Sallie, who is now seventy-six years old and seldom saw her father, saw the silhouettes and thought: "All these years I have been lied to. I was told that my father didn't like little children and that he didn't like me. But now I know that he must have loved me because he kept those silhouettes all those years."

When she told me this story, tears came to her eyes. She asked if I knew where the silhouettes were because she would like to have a copy before she died. I said I would check when I returned to Hawaii.

Again years passed. The three silhouettes where finally found in an unmarked box in the archive room at Hawaii Volcanoes National Park. Copies were made and sent to Sallie.

Sallie Jaggar died three months later.

SOURCES

Literally thousands of documents were examined in the preparation of this book. They were in the form of books, letters, memos, research papers published in scientific journals, diaries, government reports, personnel files, genealogical records, newspaper and magazine articles, accounting sheets, unpublished memoirs and field notebooks. In addition, several people were interviewed who knew Professor Jaggar personally or who were at Kilauea volcano during the early years of the volcano observatory.

GENERAL REFERENCES

Bevens, Darcy, Taeko Jane Takahashi and Thomas L. Wright. *The Early Serial Publications of the Hawaiian Volcano Observatory, 3 volumes.* Hawaii National Park: Hawaii Natural History Association. 3,062 pp. 1988.

Brigham, William T. *The Volcanoes of Kilauea and Mauna Loa on the Island of Hawaii.* Honolulu: Bishop Museum Press. 222 pp. 1909.

Fiske, Richard S., Tom Simkin and Elizabeth Nielsen. *The Volcano Letter.* Washington, D.C.: Smithsonian Institution Press, 1987.

Hitchcock, Charles H. *Hawaii and its Volcanoes.* Honolulu: Hawaiian Gazette Company. 314 pp. 1909.

Jaggar, Thomas A. *Origin and Development of Craters. The Geological Society of America. Memoir 21.* 508 pp. 1947.

Jaggar, Thomas A. *My Experiments with Volcanoes.* Honolulu: Hawaiian Volcano Research Association, Advertising Publishing Company. 198 pp. 1956.

Poland, Michael P., Taeko Jane Takahashi, and Claire M. Landowski (eds.). *Characteristics of Hawaiian Volcanoes: U.S. Geological Survey Professional Paper 1801*. 429 pp. 2014.

Wright, Thomas L. and Taeko Jane Takahashi. *Observations and Interpretations of Hawaiian Volcanism and Seismicity 1779–1955: An Annotated Bibliography and Subject Index*. Honolulu: University of Hawaii Press. 270 pp. 1989.

ARCHIVES AND COLLECTIONS

Bishop Museum, Honolulu, Hawaii
California Institute of Technology, archives, Pasadena, California
California State Archives, Berkeley, California; Sacramento, California
California State Library, California History Room, Sacramento, California
Connecticut Valley Historical Museum, Springfield, Massachusetts
Hamilton Library, University of Hawaii, archives, Honolulu, Hawaii
Hawaiian Volcano Observatory, Hawaii Volcanoes National Park, Hawaii
Hawaii Volcanoes National Park, archives, Hawaii
Massachusetts Institute of Technology, archives, Cambridge, Massachusetts
Sutro Library, San Francisco State University, genealogical library, San Francisco, California
United States Geological Survey, archives, Denver, Colorado; Reston, Virginia
United States National Archives, San Bruno, California; College Park, Maryland; St. Louis, Missouri
Vesuvius Observatory, Naples, Italy
WWII Japanese American Internment and Relocation Records, College Park, Maryland

INTERVIEWS

Jane Christman Albritton—stenographer and typist at Hawaii National Park, 1934–1936
Russell Apple—ranger and historian at Hawaii National Park from 1950
Alfred Tai On Au—machinist at Hawaiian Volcano Observatory, 1928–1933
Ernest Cabrinha—Hilo resident
Northup Castle—summertime resident at the Volcano House, 1915–1919
Gordon Cran—manager of Kapapala ranch near Hawaii National Park
Harvey Finch—Ruy Finch's son
Barbara Fitzgibbons—the Jaggars stayed at the Fitzgibbons' house when they visited Kona
John Forbes—machinist at Hawaiian Volcano Observatory, 1950–1983
Sherwood Greenwell—manager of Kealakekua ranch in Kona, west Hawaii
Eliza Bowne "Sallie" Jaggar Hayes—Thomas Jaggar's daughter
Samuel Lamb—biologist and ranger at Hawaii National Park, 1934–1938
Alexander Alika Lancaster—grandson of volcano guide Alexander Lancaster
Harold Luscomb, Jr.—Hilo resident

Abigail "Api" Kanakaole Oliveira—resident of Kapapala ranch
Akira Yamamoto—clerk and field assistant at Hawaiian Volcano Observatory, 1955–1983
Shizuka Yasunaka—Jaggars' housekeeper, 1917–1940

NEWSPAPERS AND MAGAZINES
The Boston Globe. Boston, Massachusetts
The Brooklyn Eagle. Brooklyn, New York
Daily Post-Herald. Hilo, Hawaii
Daily Tribune. Hilo, Hawaii
The Friend. Honolulu, Hawaii
Harvard Crimson. Cambridge, Massachusetts
Hawaii Herald. Hilo, Hawaii
Hawaiian Post. Hilo, Hawaii
Hawaiian Star. Honolulu, Hawaii
Hawaii Tribune-Herald. Hilo, Hawaii
Kohala Midget. Kohala, Hawaii
The New York Times. New York, New York
Pacific Commercial Advertiser. Honolulu, Hawaii
Paradise of the Pacific. Honolulu, Hawaii
San Francisco Chronicle. San Francisco, California
The Springfield Republican. Springfield, Massachusetts
Star-Bulletin. Honolulu, Hawaii
The Tech. Boston, Massachusetts
The Tribune. Hilo, Hawaii.
The Washington Post. Washington, D.C.

SELECTED SOURCES
Prologue: A City has Perished
Anonymous. "Cause of eruptions," interview of Harvard Professor Nathaniel Shaler, *Boston Globe*. p. 2. May 12, 1902.
Anonymous. "Hero of the Roddam in England," interview of Captain Edward Freeman, *London Daily Express*. June 19, 1902.
Morris, Charles. *The Destruction of St. Pierre and St. Vincent*. Philadelphia: American Book and Bible House. 432 pp. 1902.

Chapter 1: The Bishop's Son
Ames, Robert. *Official Report of the Relief Furnished to the Ohio River Flood Sufferers*. Evansville, Indiana: Journal Company. 75 pp. 1884.
Johnston-Lavis, Henry J. "Notes on Vesuvius from February 4 to August 7, 1886," *Nature*. vol. 34, p. 557-558. October 7, 1886.
Kinsolving, Rev. Arthur B. *Texas George: The Life of George Herbert Kinsolving, Bishop of Texas, 1892–1928*. Milwaukee: Morehouse Publishing Co. 137 pp. 1932.

Manross, William Wilson. *A History of the American Episcopal Church*. New York: Morehouse Publishing Co. 456 pp. 1935.

Chapter 2: Yellowstone

Blevins, Bruce H. *Absaroka Mountains 1893 and 1897: Jaggar's Diaries and Photographs*. Powell, Wyoming: WIM Marketing. 108 pp. 2002.

Jaggar, Thomas A. "Death Gulch, a natural bear trap," *Popular Science Monthly*. vol. 54., p. 475-480. February 1899.

Livingstone, David N. *Nathaniel Southgate Shaler and the Culture of American Science*. Tuscaloosa, Alabama: The University of Alabama Press. 395 pp. 1987.

Morison, Samuel Eliot. *Three Centuries of Harvard, 1636–1908*. Cambridge, Massachusetts: The Belkap Press of Harvard University Press. 520 pp. 1965.

Smith, Richard Norton, *The Harvard Century: The Making of a University to a Nation*. New York: Simon and Schuster. 397 pp. 1986.

Chapter 3: The Caribbean

Garesché, William A. *Complete Story of the Martinique and St. Vincent Horrors*. L.G. Stahl. 462 pp. 1902.

Jaggar, Thomas A. "Field notes of a geologist in Martinique and St. Vincent," *Popular Science Monthly*. vol. 61, p. 352-358. August 1902.

Jaggar, Thomas A. "Crater of Soufrière, St. Vincent," *Harper's Weekly*. vol. 46, p. 1281. September 13, 1902.

Kennan, George. *The Tragedy of Pelee*. New York: The Outlook Company. 257 pp. 1902.

Morgan, Peter. *Fire Mountain*. New York: Bloomsberg. 244 pp. 2003.

Chapter 4: Champagne

Ainsworth, William. "Notice of the volcanic island thrown up between Pantellaria and Sciacca," *American Journal of Science*. vol. 21, p. 399-404. 1832.

Anderson, Tempest and John Flett. "Report on the Eruptions of the Soufrière, in St. Vincent, and a visit to Montagne Pelee, in Martinique," *Philosophical Transactions of the Royal Society of London, Series A*. vol. 200, p. 353-553. 1903.

Hooker, Marjorie. "The origin of the volcanological concept Nuée ardente," *Isis*. vol. 56, no. 186. p. 401-407. 1965.

Jaggar, Thomas A. "Eruption of Pelee, July 9, 1902," *Popular Science Monthly*. vol. 64, p. 219-231. January 1904.

Jaggar, Thomas A. "The initial stages of the spine on Pelee," *Journal of Science*. vol. XVII, p. 34-40. January 1904.

Lacroix, Alfred. *La Montagne Pelee après ses Éruptions*. Paris: L'Académie des Sciences. 136 pp. 1908.

Chapter 5: Vesuvius

Anonymous. "The Perret automobile and battery," *The Electrical Age*. vol. 25, no. 16, p. 121. April 21, 1900.

Jaggar, Thomas A. "Eruption of Mount Vesuvius, The April 7–8, 1906," *National Geographic Magazine*. vol. 17, p. 318-325. June 1906.

Lobley, J. Logan. *Mount Vesuvius*. London: Roper and Drowley. 400 pp. 1889.

Perret, Frank. *The Vesuvius Eruption of 1906*. Washington, D.C.: Carnegie Institution of Washington. 151 pp. 1924.

Chapter 6: Alaska

Jaggar, Thomas A. "Journal of the Technology Expedition to the Aleutian Islands, 1907," *Technology Review*. Boston, Massachusetts Institute of Technology, Alumni Association. vol. 10, no. 1, 37 pp. January 1908.

Jaggar, Thomas A. "The evolution of Bogoslof volcano," *Bulletin of the American Geographical Society*. vol. 40, no. 7, p. 385-400. 1908.

Seeley, George. *Manifest of the Ship "Lydia."* Available from FamilySearch.org microfilm no. 1666641. 1907.

Chapter 7: The Pacific World

Bevens, Darcy. *On the Rim of Kilauea: Excerpts from the Volcano House Register 1865–1955*. Hawaii National Park: Hawaii Natural History Association. 168 pp. 1992.

George, Milton C. *The Development of Hilo, Hawaii, T.H.: A Modern Tropical Sugar Port*. Ann Arbor, Michigan: The Edwards Letter Shop. 62 pp. 1948.

Jaggar, Thomas A. "Japanese volcanoes," *Bulletin of the Society of Arts*. February 1910.

Kelly, Marion, Barry Nakamura, and Dorothy Barrere. *Hilo Bay, A Chronological History*. Prepared for U.S. Army Engineer District, Honolulu. 341 pp. 1981.

Reid, Harry Fielding. "On the choice of a seismograph," *Bulletin of the Seismological Society of America*. vol. 2, no. 1, p. 8-20. March 1912.

Strzelecki, Paul Edmund. "Sandwich Islands—Crater of Kirauea, Hawaii," *Hawaiian Spectator*. vol. 1, p. 434-437. October 1838.

Chapter 8: Into the Cauldron

Bonney, Thomas. *The Story of Our Planet*. London: Cassell & Company. 535 pp. 1893.

Geikie, Archibald. *Text-book of Geology*. London: Macmillan and Company. 992 pp. 1885.

Jaggar, Thomas A. "The earthquake in Costa Rica," *Science Conspectus*. vol. 1, no. 2, p. 33-40. January 1911.

Judd, John W. *Volcanoes: What They Are and What They Teach*. London: Kegan Paul, Trench & Company. 381 pp. 1881.

Perret, Frank. "Volcanic research at Kilauea in the summer of 1911," *American Journal of Science*. vol. 36, Fourth Series, p. 474-483. November 1913.

Sigurdsson, Haraldur. *Melting the Earth: The History of Ideas on Volcanic Eruptions*. New York: Oxford University Press. 272 pp. 1999.

Stevens, Sylvester K. *American Expansion in Hawaii 1842–1898*. Harrisburg: Archives Publishing Company of Pennsylvania. 320 pp. 1945.

Thurston, Lorrin A. *Writings of Lorrin A. Thurston*. Honolulu: Advertiser Publishing Company. 168 pp. 1936.

Westervelt, William D. "Scientists and Madame Pele," *Paradise of the Pacific*. September 1911.

Chapter 9: A Dream Fulfilled

Anonymous. "The Fires of Kilauea Reproduced," review of the play "Bird of Paradise," *Honolulu Commercial Advertiser*. p. 3. June 16, 1912.

Brun, Albert. *Recherches sur l'Exhalaison Volcanique*. Paris: A. Hermann & Fils. 277 pp. 1911.

Day, Arthur L. and Ernest S. Shepherd. "Water and volcanic activity," *Bulletin of the Geological Society of America*. vol. 24, p. 573-606. 1913.

Dodge, Frank B. *The Saga of Frank B. Dodge: An Autobiography*. Tucson, Arizona: U.S. Geological Survey Administrative Report. 1944.

Jaggar, Thomas A. "The Technology Station in Hawaii," *Technology Review*. Boston, Massachusetts Institute of Technology, Alumni Association. vol. XIII, no. 8. November 1911.

Chapter 10: A Love Lost

Jaggar, Thomas A. "Activity of Mauna Loa, Hawaii, December–January, 1914–1915," *American Journal of Science*. vol. 40, p. 621-639. December 1915.

Jaggar, Thomas A. "Notes from a volcano laboratory," *Scientific American, Supplement No. 2074*. p. 214-217. October 2, 1915.

Jaggar, Thomas. *A Bill [H.R. 9525] to Establish a National Park in the Territory of Hawaii: Hearing before the Committee on Public Lands, February 3, 1916*. Washington, D.C.: Government Printing Office. p. 3-30. 1916.

Jaggar, Thomas A. "Sakurajima, Japan's greatest volcanic eruption," *National Geographic Magazine*. vol. 24, no. 4. April 1924.

Smits, Gregory. *When the Earth Roars*. Lanham, Maryland: Rowman & Littlefield. 209 pp. 2014.

Chapter 11: The School Teacher

Jaggar, Thomas A. "Lava flow from Mauna Loa, 1916," *American Journal of Science, Fourth Series*. p. 256-288. April 1917.

Withington, Antoinette. *Hawaiian Tapestry*. New York: Harper & Brothers. 367 pp. 1937.

Wood, Harry Oscar. "Notes on the 1916 eruption of Mauna Loa," *Journal of Geology*. vol. 25, p. 322-336. 1917.

Yardley, Paul T. *Millstones and Milestones: The Career of B.F. Dillingham, 1844–1918*. Honolulu: University of Hawaii Press. 339 pp. 1981.

Chapter 12: The Lava Lake

Brown, Ernest W. "Tidal oscillations in Halemaumau, the pit of Kilauea," *American Journal of Science*. Fifth Series. vol. 9, no. 2, p. 95-112. February 1925.

Coan, Titus. "Great eruption of the volcano of Kilauea," *Missionary Herald*. p. 283-285. July 1841.

Ellis, William. *A Journal of a Tour Around Hawaii*. New York: Crocker & Brewster. 264 pp. 1825.

Jaggar, Thomas A. "Volcanologic investigations of Kilauea," *American Journal of Science. Fourth Series*. vol. 44, p. 161-220. September 1917.

Jaggar, Thomas A. "Seismometric investigation of the Hawaiian lava column," *Bulletin of the Seismological Society of America*. vol. 10, no. 4, p. 155-275. December 1920.

Jaggar, Thomas and Ruy H. Finch. "The explosive eruption of Kilauea in Hawaii, 1924," *American Journal of Science. Fifth Series*. vol. 8, no. 47, p. 353-374. November 1924.

Jaggar, Thomas A., Ruy H. Finch and Oliver Emerson. "The lava tide, seasonal tilt, and the volcano cycle," *Monthly Weather Review*. vol. 52, p. 142-145. March 1924.

Wilkes, Charles. *Narrative of the U.S. Exploring Expedition During the Years 1838–1842*. 5 vols., Philadelphia: Lea and Blanchard. 1845.

Chapter 13: Mauna Loa

Anonymous. "Tales of Madame Pele on Kona side of island held omen of volcano activity," *Hilo Tribune-Herald*. p. 1. March 7, 1926.

Anonymous. "Woman is attacked by shark," *Hawaii Tribune-Herald*. p. 1. April 8, 1926.

Anonymous. "Madame Pele betrays trust," *Pacific Commercial Advertiser*. p. 1. April 20, 1926.

Dutton, Clarence. "The Hawaiian volcanoes," *U.S. Geological Survey, 4th Annual Report*. Washington, D.C.: Government Printing Office. p. 75-219. 1884.

Jaggar, Thomas A. "Experiences in a volcano observatory," *Natural History*. vol. 21, no. 4, p. 337-342. July-August 1921.

Ledyard, John. *A Journal of Captain Cook's Last Voyage to the Pacific Ocean and in Quest of a Northwest Passage between Asia and America*. Hanford, Connecticut: Nathaniel Patten. 208 pp. 1783.

Menzies, Archibald. *Hawaii Nei 128 Years Ago*. Honolulu: W.F. Wilson. 199 pp. 1920.

Chapter 14: The Goddess

Emerson, Nathaniel B. *Pele and Hiiaka: A Myth from Hawaii*. Honolulu: Honolulu Star-Bulletin. 250 pp. 1915.

Fornander, Alexander. *Hawaiian Antiquities and Folk-lore*. Honolulu: Bishop Museum Press. vol. IV, part II. 1919.

Gregg, C.E. and others. "Hawaiian cultural influences on support for lava flow hazard mitigation measures during the January 1960 eruption of Kilauea volcano, Kapoho, Hawaii," *Journal of Volcanology and Geothermal Research*. vol. 172, p. 300-307. 2008.

Handy, E.S. Craighill and Mary Kawena Pukui. *The Polynesian Family System in Ka'u, Hawai'i*. Rutland, Vermont: Charles E. Tuttle Co. 259 pp. 1972.

Jaggar, Thomas A. "Protection of harbors from lava flow," *American Journal of Science.* vol. 243-A, p. 333-351. 1945.

Kalakaua, David. *The Legends and Myths of Hawaii.* New York: Charles L. Webster & Company. 530 pp. 1888.

Nimmo, H. Arlo. *Pele, Volcano Goddess of Hawai'i: A History.* Jefferson, North Carolina: McFarland & Company. 239 pp. 2011.

Pukui, M.K., E.W. Haertig, and C.A. Lee. *Nana i ke kumu (Look to the Source).* 2 vols. Honolulu: Queen Liliuokalani Children's Center Publication. 1972–1979.

Swanson, Donald A. "Hawaiian oral tradition describes 400 years of volcanic activity at Kilauea," *Journal of Volcanology and Geothermal Research.* vol. 176, Westervelt, William D. *Hawaiian Legends of Volcanoes.* Boston: G.H. Ellis Press. 205 pp. 1916.

Chapter 15: The Last Volcano

Finch, Ruy H. "On the prediction of tidal waves," *Monthly Weather Review.* vol. 52, p. 147-148. March 1924.

Jaggar, Thomas A. "Mapping the home of the great brown bear," *National Geographic Magazine.* vol. 55, p. 109-134. January 1929.

Jaggar, Thomas A. "The Great Tidal Wave of 1946," *Natural History.* vol. 55, no. 6, p. 263-268, 293. 1946.

Smits, Gregory. *When the Earth Roars.* Lanham, Maryland: Rowman & Littleman. 208 pp. 2014.

Chapter 16: A Forgotten Legacy

Anonymous. "Invention of 'Duck' wins prize for Dr. Jaggar," *Science News-Letter.* vol. 47, no. 9, p. 134. March 3, 1945.

Gerlach, Terry. "Volcanic versus anthropogenic carbon dioxide," *EOS, Transactions of the American Geophysical Union.* vol. 92, no. 24, pp. 201-208. June 14, 2011.

Jaggar, Thomas A. *Volcanoes Declare War: Logistics and Strategy of Pacific Volcano Science.* Honolulu: Paradise of the Pacific. 166 pp. 1945.

Kashima, Tetsuden. *Personal Justice Denied: Report of the Commission on Wartime Relocation and Internment of Civilians.* Seattle: University of Washington Press. 493 pp. 1997.

Ogawa, Dennis and Evarts Fox. "Japanese interment and relocation: The Hawaii experience," in Daniels, Roger, Sandra Taylor and Harry Kitano, eds. *Japanese Americans: From Relocation to Redress.* Salt Lake City: University of Utah Press. p. 135-138. 1986.

INDEX